Crayfish, Crawfish, Crawdad

# CRAYFISH,

# Crawfish,

# CRAWDAD

## The Biology
## and Conservation of
## North America's
## Favorite Crustaceans

## Zackary A. Graham

The University of North Carolina Press
CHAPEL HILL

This book was published with the assistance of the Blythe Family Fund of the University of North Carolina Press.

Designed by Lindsay Starr
Set in Quadraat and IM Fell Double Pica
by Rebecca Evans

Cover art: A Depression Crayfish (*Cambarus rusticiformis*).
Photo by Guenter Schuster.

Library of Congress Cataloging-in-Publication Data
Names: Graham, Zackary A. author
Title: Crayfish, crawfish, crawdad : the biology and conservation
of North America's favorite crustaceans / Zackary A. Graham.
Description: Chapel Hill : University of North Carolina Press,
[2026] | Includes bibliographical references and index.
Identifiers: LCCN 2025029895 | ISBN 9781469685731 paperback
alk. paper | ISBN 9781469685748 epub | ISBN 9781469685755 pdf
Subjects: LCSH: Crayfish—United States—Popular works | Crayfish—
Ecology—Popular works | Crayfish—Conservation—Popular works |
Endangered species—United States | BISAC: NATURE / Animals /
General | SCIENCE / Life Sciences / Zoology / Invertebrates
Classification: LCC QL444.M33 G73 2026
LC record available at https://lccn.loc.gov/2025029895

For product safety concerns under the European Union's General Product Safety
Regulation (EU GPSR), please contact gpsr@mare-nostrum.co.uk or write to
the University of North Carolina Press and Mare Nostrum Group B.V.,
Mauritskade 21D, 1091 GC Amsterdam, The Netherlands.

# Contents

Chapter 1. An Introduction to Crayfish · 1

Chapter 2. Mystery of the Mudbugs · 17

Chapter 3. Ecosystem Engineers and Keystone Species · 45

Chapter 4. Crayfish Brains and Crayfish Benders · 68

Chapter 5. Communicating with Urine · 86

Chapter 6. Aquatic Invaders · 116

Chapter 7. *Pacifastacus*: The Crayfishes of the Pacific Northwest · 146

Chapter 8. *Cambarus callainus* and *Cambarus veteranus*:
The Crayfish in the Coalfields · 169

Chapter 9. Describing (and Conserving) the Future of Crayfish · 191

ACKNOWLEDGMENTS · 203

BIBLIOGRAPHY · 205

INDEX · 215

# Crayfish, Crawfish, Crawdad

# An Introduction to Crayfish

## Flipping Rocks

I'VE LOVED COLLECTING and learning about animals for as long as I can remember. Unlike the other neighborhood children who sat outside on warm summer days trying to guilt the neighbors into buying lemonade, I was out there selling bugs—dead bugs. Beetles, crickets, butterflies, cicadas, you name it. My hand-caught inventory was neatly organized in the compartments of an old-school tackle box propped up on a white folding table with a large sign in front: BUGS FOR SALE. Despite my determination, I only made a few quarters from this endeavor. Nearly ten years later, I found out that my sales were not because of my impressive inventory or customer service skills but rather the result of my mom furiously calling the neighbors and pleading with them to support my business.

One of my later animal escapades involved my best friend and me (who, amusingly, is named Zach) traversing a narrow backyard valley to a stream we called "the ravine," flipping rocks and logs the whole way down. Every animal we found, we collected for later identification. Old Tupperware containers became habitats for our collections, which ended up taking over a whole corner of Zach's parents' garage. One time, we thought we had bred salamanders in one of our enclosures, only to learn that the little wiggly worms we thought were salamander babies were actually thousands of mosquito larvae, a much less impressive feat.

Aside from local adventures, most of my animal obsessions came from a yearly stay at the Linn Run State Park rustic cabins in the foothills of the Allegheny Mountains. Staying at Linn Run was a tradition my family held the entirety of my grade school years. While only sixty miles outside of Pittsburgh, Pennsylvania, this park felt like a different world. Spring-fed streams toppled down the mountains into Linn Runn, creating a pristine aquatic habitat. When I was at the cabin, from sunrise to sunset my days were spent outside looking for frogs, salamanders, and my personal favorite, crayfish. At night, I would try to match the animals I found with pictures in the Audubon Society field guides. Inspired by nature educators like Steve Irwin and Jeff Corwin, I imagined I was in an undiscovered land trying to identify animals that rarely ventured into my usual suburban habitat. The rush of dopamine that I got (and still get) from flipping a rock and seeing what's underneath is one of the best feelings in the world. Catching and learning about animals makes me feel better than anything else I have pursued. And I can't remember a time when I haven't felt that way.

Although crayfish are generally an easy target for animal catchers, there are always exceptions to the rule. At Linn Run, most of the rocks are shaped like a dinner plate, which means I could easily squeeze my little fingers underneath them to search for crayfish. But some of the rocks are shaped more like bowling balls and become completely embedded in the stream bank. Only a few years after learning to tie my shoes, I remember trying to wedge my fingers under one of these bowling-ball rocks that was nearly the size of my torso—but without success. I ended up taking a different route. Instead of trying to muster up the strength to flip the rock, I dug underneath it and toppled the miniature boulder into the water.

The largest rocks often hold the largest surprises (and now that I am older, the largest amount of back pain), and the muddy water eventually cleared up to reveal a large white crayfish. It had nowhere to run, and I quickly grabbed the milky-white, almost translucent crayfish, which fit perfectly in my hand—the largest I had ever seen. Having collected hundreds of crayfish in this stream, I had never seen one quite like it. But aside from its ghostly color, it looked nearly identical to the typical brown and tan species I was familiar with. Knowing about color mutations like albinism and melanism, I assumed this was an albino crayfish, because its color was nothing like that of the typical brownish, rust-colored species I was used to. Now, as an adult, and remembering that this crayfish had black eyes, I know that the scientific term for such a crayfish is "leucistic" (primarily white, but not completely). And thousands of rock flips later, I have yet to find another specimen quite like it. The dopamine rush was strong that day.

As a teenager, animals were always at the back of my mind, but my priorities changed. I focused on sports like volleyball and swimming. Becoming a biologist (and especially a crayfish biologist) was never on my radar. I eventually did find my way back into biology through undergraduate research experiences that allowed me to study both human mate preferences and lizard brains with two different professors. These experiences got me back on my animal kick, and I haven't stopped since. I went straight from my undergraduate studies to pursue a PhD in animal behavior at Arizona State University, one of few programs of its kind in the United States. At Arizona State, I was given free rein to choose my research topic. I am sure you can guess what organism I ended up studying. I'll cover my PhD research in a later chapter, but as a teaser, it involves sex, fighting, and urine. I was unaware that this decision would impact my life so much, but I was just chasing the same rush of dopamine I got as a kid.

During my PhD, I began learning about the complexities of crayfish. I knew the basics from my childhood of catching them, but it takes years to get your feet wet in crayfish biology, or astacology. Fundamental information on these animals is buried in books spanning hundreds of pages—something that few ever have the fortitude (or time) to tackle. By reading these texts and more recent academic work, I learned that crayfish don't inhabit just streams. I learned that they don't come in just camouflaged earth-toned colors. I learned that there are more species in North America than anywhere else in the world. And most of all, I learned that they are among the most threatened groups of animals in the world.

These animals are so charismatic and quirky, with almost everyone having some experience with them at some point in their life. But there is almost no way for nonexperts to learn about them. Early on my journey to becoming a crayfish biologist, while learning to differentiate the species I was catching in Pennsylvania, I was daunted by the descriptions of how to tell them apart: "First form gonopods short, comprising 33% of TCL, with two terminal elements about equal length; corneous central projection comprising 23%

of pleopod length, tapering distally to point; mesial process non-corneous, spatulate, partially surrounding central projection; cephalic base of central projection with right angle shoulder." The initial barrier encountered by most nonexperts is usually the new vocabulary, which can be daunting and takes some focused study and dedication to overcome. I hope to knock down this barrier so that learning about crayfish is more of a crawl than a climb.

I HAVE THREE GOALS in writing this book. First, I want to show that crayfish are more than a boring brown bug that lives under rocks in the water. Crayfish have adapted to nearly every freshwater environment imaginable. Second, I want to show that crayfish are ecologically important, and without them, North American ecosystems would not be the same. Third, I want to show that crayfish are facing an extinction crisis. Half of all North American species require conservation attention, and without intervention many will continue to decline and disappear.

The crayfish stories will be told through my eyes and through the eyes of other crayfish biologists who, like me, have never lost that spark that comes from catching and learning about animals. This journey will follow the paths of the crayfish from the coalfields of Central Appalachia to the spring-fed water of Northern California and roadside ditches everywhere in between.

## What Is a Crayfish?

One of the biggest issues with raising awareness for crayfish is just how many names there are for these animals. You may be wondering, Is a crayfish the same thing as a crawfish? Is it the same thing as a crawdad? Or a mudbug? The answer is yes. Crayfish, crawfish, crawdad, and mudbug are all common names for the same animal. Like soda, pop, and coke, they are all names for the same thing—just different local variations. Often, what you call these animals entirely depends on where you are from. If you are from the Northeastern United States or the Great Lakes region, you likely call them crayfish. If you are from the Midwest, you likely call these animals crawdads. Almost everywhere else, including the Southern United States, they are called crawfish. Growing up in Pennsylvania, I always knew these animals as crayfish, but when I lived in Arizona, I learned that the locals call them crawdads. Still, a few pockets of the United States have derived their own unique names for them, like pond lobsters, ditch crickets, or crawcrabs.

Within the world of crayfish biology, these animals are formally called crayfish. There's a saying that I have heard over and over in the crayfish community: "If you are studying them, call them crayfish; if you are using

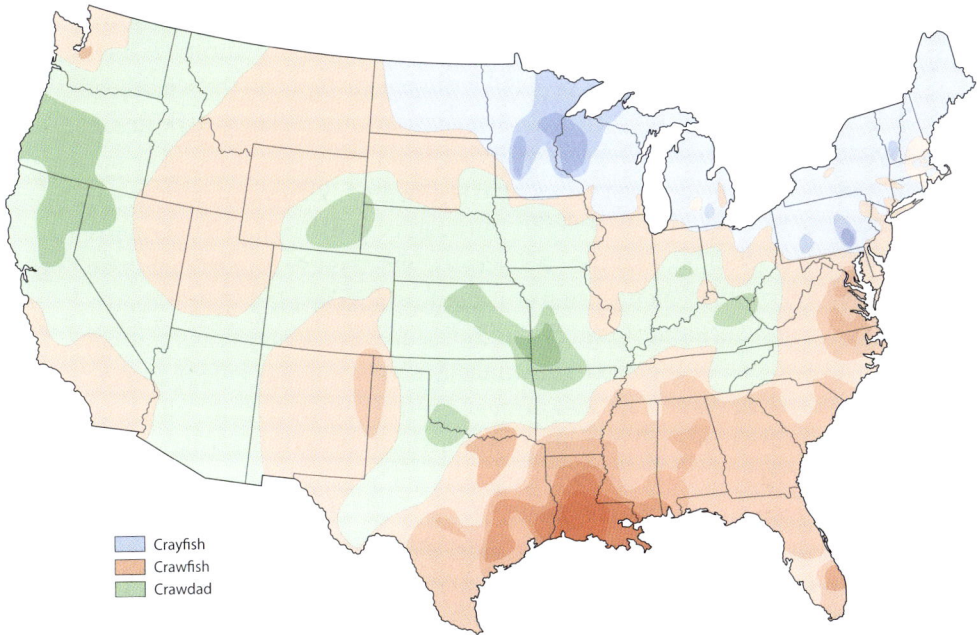

A visual heat map of survey responses from individuals who were asked the question "What do you call the miniature lobster that one finds in lakes and streams?" Darker colors indicate a greater percentage of the population choosing one answer over another.

them as fishing bait, call them crawdads; and if you are eating them, call them crawfish." I will refer to them throughout this book as crayfish, because that is the now more or less settled-on common name. However, in certain contexts, such as in referencing the wild harvest of these crustaceans throughout Louisiana, out of respect for their culture, I will call them crawfish, just as the locals do.

So why are they called crayfish if they are clearly not fish? Like fish, all crayfish rely on water in some way or another, but the origin of the name actually has nothing to do with them sharing the aquatic environment with fish. The French word for this animal is *écrevisse* [eh-kruh-vees], and apparently, after several misinterpretations and mutations of the word, the English translation popped out as "crayfish." Note that the term "crayfish" can refer to a single individual or multiple individuals of the same species, whereas "crayfishes" is used specifically when referencing multiple species of crayfish.

Crayfish and insects belong to a group of animals called invertebrates, which comprises around 95% of all living animals. Invertebrates lack a backbone, which differentiates them from vertebrates, such as humans, birds, and fish. Like insects and spiders, crayfish are arthropods. Arthropods have an exoskeleton, meaning that their skeleton effectively is on the outside of the body instead of on the inside. Arthropods have no soft exposed flesh like humans do, and instead they wield this hardened, waterproof suit of armor to protect their insides. They also have segmented bodies and jointed appendages, which give them flexibility. Although crayfish resemble scorpions (which are arachnids), they are crustaceans, which is a separate taxonomic grouping altogether. Noteworthy crustaceans include a lot of well-known edible seafood, including lobsters, shrimps, and crabs. Crustaceans are also known to be one of the most anatomically diverse animal groups to ever exist, with a huge variety in body form and function. Flip over a rock in any moist area anywhere in the world and you are likely to find terrestrial crustaceans: the armadillo-like isopods known as pill bugs, sow bugs, wood lice, or as my niece calls them, roly-polies.

Crayfish belong to the taxonomic order Decapoda. In Latin, *deca* means "ten" and *poda* means "feet"; all decapods have ten primary appendages. Many decapod crustaceans have evolved their most frontal appendages into highly specialized and exaggerated claws. Although crayfish and lobsters are commonly confused based on their shockingly similar body plan, they diverged evolutionarily from each other roughly 330 million years ago. Lobsters are restricted to saltwater environments, whereas crayfish are tied to freshwater. Crayfish are native to every continent except Antarctica and mainland Africa (but there are crayfish on Madagascar!). Crayfish are also more diverse than lobsters. Only thirty species of lobsters are known, but there are over 700 species of crayfish, with more and more being described each year.

Crayfish and lobsters get confused *a lot*. While wandering the fish section of pet store chains, it's funny to see colorful genetic variants of crayfish being sold under the label "Electric Blue Lobster" or "Ghost Lobster." In most scenarios, the workers are (understandably) naïve to the taxonomic differences between the two. But in other cases, their intentions may be deceptive. Intentionally labeling crayfish as lobsters is a way to get around the illegal trade of crayfish in some areas—a topic we'll revisit later.

Why do crayfish and lobsters look so similar? It is a classic evolutionary case of "If it ain't broke, don't fix it." The body plans of crayfish and lobsters seem to have filled a similar ecological role in both freshwater and saltwater environments. Their bodies are heavily coated with chitin, the same material that comprises all arthropod exoskeletons, which offers protection from predators. Crayfish and lobsters also both have a fused head and body

# CLASSIFYING CRAYFISH

All animals fit into a biological filing system known as taxonomy. This system allows biologists to organize and separate organisms from one another based on their similarities or differences. Below are the most important taxonomic classifications for the most widely consumed crayfish in the world, the Red Swamp Crayfish (*Procambarus clarkii*). Following each taxonomic rank below is a short description of the major traits of each taxonomic grouping.

Kingdom: **Animalia**—the animals

Phylum: **Arthropoda**—the arthropods, a group of animals commonly called "bugs," which contains insects, arachnids (spiders, scorpions), crustaceans, and a few other groups

Subphylum: **Crustacea**—the crustaceans, a subset of arthropods, many of which (but not all) live in association with water

Class: **Malacostraca**—a group of crustaceans known best for their edibility: crawfish, shrimps, lobsters, and crabs all belong to this group

Order: **Decapoda**—meaning "ten legs," the members of the order Decapoda have ten primary appendages

Infraorder: **Astacidea**—a group that contains the freshwater crayfish, as well as their closest living relatives, clawed lobsters

Family: **Cambaridae**—the primary family of American crayfish; all Cambaridae crayfish also exhibit reproductive form alteration, which means they switch between times of being able to reproduce and times of being unable to reproduce

Genus: *Procambarus*

Species: *clarkii*

As noted above, all crayfish in the family Cambaridae are native to the Americas. Below the family level, different crayfish are also assigned to a genus and to a species, which, when stated together, is known as their scientific name (e.g., *Procambarus clarkii*). Throughout this book, I will refer to many crayfish by their scientific name, although each species also has a common name (e.g., the Red Swamp Crayfish). The scientific names may seem intimidating at first, but a scientific name has several benefits over the common name. Primarily, common names are often of an unspecific nature and lack a standardized system. For example, the common name of one of the most widespread species of crayfish in North America is the Common Crayfish—with the scientific name of *Cambarus bartonii*. This common name can be extremely misleading, because depending on where you live, the Common Crayfish may be rare or, in some cases, nonexistent. Therefore, scientific names are more descriptive and are preferred in many cases. Further complicating things, one species may have several common names in use. The Common Crayfish (*Cambarus bartonii*) is sometimes referred to ask the Eastern Crayfish or the Appalachian Brook Crayfish.

You will know when the scientific name is being discussed based on the standard way such names are displayed: The first letter of the genus name is capitalized and the species name is lowercase, and both names together are italicized (e.g., *Procambarus clarkii*, *Cambarus bartonii*). Typically, I list a species common name in plain text immediately before giving its scientific name (e.g., the Ditch Fencing Crayfish, *Faxonella clypeata*). In an attempt to highlight the animals themselves, I present common names as if they are proper nouns. After the first use of a species' scientific name, the genus name can be abbreviated to avoid repetition of lengthy names while maintaining clarity (e.g., *P. clarkii*, *C. bartonii*, *F. clypeata*).

(thorax), forming what's called a cephalothorax. But the structure that humans are most familiar with in both animals is their abdomen, which is responsible for creating the iconic tail-flip retreat and also where the meat that humans consume comes from. Crayfish and their lobster cousins can create one strong tail thrust and quickly dart away from predators (this behavior is sometimes called crawfishing, and "to crawfish" can mean to quickly back away from something). Moreover, in both crayfish and lobsters, both sexes wield a pair of enlarged claws, which in some species can match their body size in length and make up 40% of their adult body weight. Claws are multifunctional organs and are a primary way for these animals to interact with their environment; they serve as weapons during aggression, sexual signals during mating, deterrents in predator defense, and utensils in feeding.

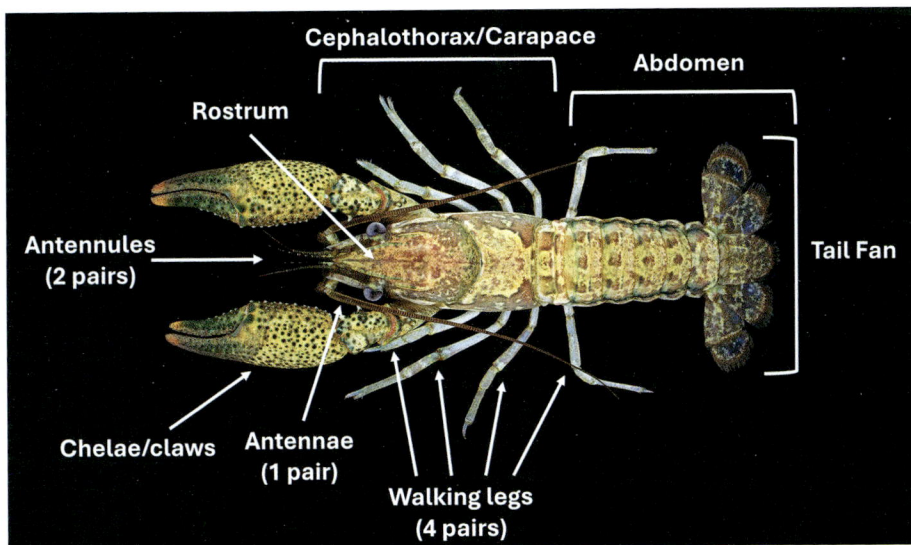

The external anatomy of a crayfish. Photo edited based on image provided by Guenter Schuster.

Being low to the ground and having thick armor and a big set of claws makes crayfish and lobsters resemble aquatic tanks. But their general body plan can adapt to different environments, which allows them to inhabit otherwise uninhabitable locations, like caves and terrestrial burrows. Since there are more than twenty times the number of crayfish species as lobster species, you can imagine that crayfish have had to survive in a variety of places, which has led to greater diversity across the board. Creating more name-based confusion, in some parts of the world, such as Australia, some crayfish species are given a common name that includes the word "lobster," and a different group of crustaceans called spiny lobsters (which are not related to standard lobsters) are given the common name "crayfish."

## Old Man Crawfish

Becoming a surgeon takes years of training, but many students conduct their first surgery during a high school or college anatomy class. But instead of a many-hour event with a human on a table, it's for forty-five minutes and with a crayfish. This is for good reason. Most crayfish are large and easy to collect and have identifiable organs, which makes them perfect for educational dissections. Back in 1931, Horton Holcombe Hobbs Jr. (one of the most epic names imaginable) began his journey to becoming a crayfish expert by slicing open the hard exterior of a crayfish's carapace in his freshman anatomy class at the University of Florida. Initially, he was frustrated after only finding indiscernible goo underneath the hard exoskeleton. He wanted to give it another shot, but he didn't want to wait until his next class. He wanted immediate redemption.

At the time, Hobbs didn't know it, but the streams, rivers, creeks, swamps, caves, and muddy roadside areas in his home state of Florida housed nearly sixty different crayfish species. He searched his nearby environs for candidates for extrasurgical practice to take back to his dorm room. A half hour and a few rock flips later, Hobbs had collected some surgery victims. After a few practice runs in a musty-smelling dorm room, he was left with two extra crayfish, which he kept as pets in a small aquarium. At the time, he knew nothing about crayfish biology, but he enjoyed watching the animals interact and wedge themselves into the aquarium gravel to create small hiding spots. Two weeks later, Hobbs witnessed his two crayfish giving each other a face-to-face hug that lasted several minutes. Due to the lingering smell from Hobbs's crayfish dissections, I imagine this was the first time that sex occurred in that dorm room. Like humans, crayfish are one of the only animals that have sex facing each other.

Hobbs had unknowingly stuck a male and a female crayfish together during their mating season, a rookie mistake. The embrace was bound to happen. Hobbs must have been in class and sadly missed out on the pre-sex fight that crayfish engage in before they copulate. Gentle claw taps turn into full-fledged combat, which ends with the male and female clasped together. Fortunately, Hobbs did get to see the outcome of this dorm-room hookup. Three weeks after the embrace, the female extruded hundreds of eggs, which she coddled beneath her abdomen for the first few weeks of life. Hobbs surely became overwhelmed with this new responsibility but was amazed how easy it had been to breed this species and keep them in his dorm room. Like many others who have such incidental experiences, Hobbs immediately became obsessed with learning about crayfish biology. He was most interested in discovering what types of crayfish were around him. This task didn't take long, because at the time, our understanding of US (and Florida) crayfish was almost nonexistent. Later, Hobbs was given a small amount of funding to start his scientific career studying the crayfish of Florida after convincing his university dean that it was a worthy endeavor. I am willing to bet the dean had a soft spot for crayfish—maybe childhood memories of catching them.

Fast-forward through his decorated career, and Horton H. Hobbs Jr. had become synonymous with astacology and earned himself several iconic titles from his peers, including "Old Man Crawfish," "Crawdaddy," and "The Great White Crawfish." Hobbs was also a proponent of unifying the species under the name "crayfish," as earlier work used different names for the animals, which created unnecessary confusion. In total, Hobbs described 168 crayfish species and he wrote extensively about crayfish taxonomy, evolution, and biogeography. And if that wasn't impressive enough, he also described 104 ostracods (commonly known as seed shrimps), which are another crustacean species, known to live in the gill chamber of crayfish. Hobbs collected and recorded somewhere in the ballpark of 100,000 crayfish specimens during his lifetime. This collection is now housed at the Smithsonian National Museum of Natural History in Washington, DC, where Hobbs spent most of his academic career. Hobbs passed away from heart disease on March 22, 1994, at the age of ninety. Hobbs's experiences and achievements

Dr. Horton Holcombe Hobbs Jr. examining a live crayfish in his office at the US National Museum (now the Smithsonian National Museum of Natural History) in Washington, DC. Hobbs was a curator in the Department of Zoology there for twenty-seven years. Photo courtesy of Smithsonian Institution Archives, image no. 82-11578-08.

outlined above come from Karen Reed and Raymond B. Manning, both recognized crustacean biologists, who published an obituary and biographical note for Hobbs in the *Proceedings of the Biological Society of Washington*, a journal that had published Hobbs's work dozens of times throughout his career.

The godfather of modern crayfish biology and I were only alive at the same time for nineteen days, but he and I share a common story of becoming infatuated with crayfish early in life.

## More Than Just a Brown Bug

Most people are only familiar with crayfish that are earth toned in color: brown, tan, olive, or gray. In most environments, this color is probably adaptive, because it allows the crayfish to blend in with their benthic (bottom-dwelling) habitat and avoid being detected by predators. In some cases, I have waded in streams with an unusually high number of burnt-orange-colored rocks. Low and behold, the (uninformatively named) crayfish in these streams, the Rock Crayfish (*Cambarus carinirostris*), which is typically chestnut brown in color, was an unmistakable burnt orange. The pressure to hide from predators can be a strong one.

Camouflaged crayfish make plenty of sense, but other species are brilliantly colored. Crayfish can be blue, orange, red, purple, aquamarine, you name it. I am always shoving my phone in the faces of my friends, family members, and sometimes even strangers—showing off my recent catches as a proud crayfisherman. All too often, I get accused of falsifying these images, because many believe that a crayfish is "just a boring brown bug."

The aesthetic diversity doesn't stop at wild coloring. Some species lack color pigments altogether because they live in caves—appearing ghostlike with oddly shaped claws and empty eye sockets. And while many crayfish live under rocks in streams, burrowing crayfish live in extensive and intricate mud burrows that can be ten feet deep or more. In the United States, the largest species of crayfish, in the genus *Barbicambarus*, resides exclusively under rocks the size of your dinner table and can reach the size of a loaf of bread. The smallest North American crayfish—the Least Crayfish (*Cambarellus diminutus*)—is no larger than a quarter and inhabits short-lived ponds. Hopefully even just these few facts make it clear that a crayfish is more than "just a boring brown bug."

Of the more than 700 species worldwide, over 400 of them are native to the United States. Much of this diversity is in the southeastern portion of the United States, which has a long and rich history of cultural appreciation of crayfish (or as they call them there, crawfish). Historic glaciation events and the formation of the Appalachian Mountains have done wonders for the diversification of animals in this region. In Alabama alone, there are currently 100 described species; the dedicated biologists who worked on *The Crayfishes of Alabama* for over fifteen years found their progress hampered by the constant flow of new species being described each year. That's right, you don't have to go to the Amazon to find animals unknown to science. New crayfish species are being described all the time; species unknown to science are sometimes just a rock flip away.

Even more important is the fact that without crayfish, the freshwater (and terrestrial) habitats of the United States would be unrecognizable. The presence of a native crayfish community is a bioindicator of an environment's health and wellness; if native crayfish are thriving and capable of surviving,

The Least Crayfish (*Cambarellus diminutus*) is the smallest species of crayfish, reaching only 25 mm, or about 1 inch, in length as a full-grown adult. Wikimedia Commons.

The blue-and-red color morph of the Upland Burrowing Crayfish (*Cambarus dubius*). This species exhibits a wide range of colors across its range throughout Appalachia. Photo by Josh Klostermann.

so are the rest of the organisms in the stream. This is especially true for North American crayfish species, which require a narrow range of water parameters to trigger breeding and raise their young—meaning their environment has to be healthy and stable for them to thrive. Freshwater streams without crayfish are often deficient in biodiversity. If water quality diminishes due to, for example, sediment buildup from nearby road construction, the crayfish will suffer—and when crayfish suffer, so will valuable sport fish like trout and bass. At certain times of the year, sport fish rely on the protein-rich abdomens of crayfish for food. The $115 billion US recreational fishing industry is just one of many that depend on crayfish. Additionally, when crayfish disappear, aquatic plant and insect populations burst out of control, because they are not being kept in check by the crayfish's appetite for them.

NOT ONLY DO CRAYFISH serve a vital role in our environments, but they are also useful to scientists worldwide. Crayfish were first highlighted in the scientific world by Thomas Henry Huxley, nicknamed "Darwin's bulldog" for being ruthless in defending Charles Darwin's evolutionary theory. Huxley wrote a book that introduced basic zoological principles to its readers using crayfish as an example. He was also a huge proponent of pushing scientific literacy and education. Since Huxley, biologists have used crayfish to study everything under the sun—including some rather odd questions, like what happens when you give a crayfish drugs (see chapter 5).

Why are crayfish used in pursuit of scientific knowledge? As my childhood stories show, they're easily caught. Many crayfish species can be collected in the hundreds and take well to life in a lab. It is a whole lot easier to go out and catch 100 crayfish and house them in a laboratory than it is to go out and catch 100 elephants. Scientists from many disciplines use crayfish: Physiologists study how crayfish brains influence their movement patterns, geneticists study their evolutionary relationships and how they dominate freshwater environments worldwide, and behaviorists study their fighting behavior to learn what traits predict fight outcomes. Although all these fields have benefited from using crayfish as a model organism, the most abundant group of scientists studying crayfish are conservation biologists.

Soon after Hobbs and his colleagues started to fully understand the extent of crayfish diversity in the United States, they started to realize that many of these species had small distributions; some inhabit only a single stream. A new movement was born based on these realizations. Because studies have shown the unique ecological role of crayfish, scientists know that losing crayfish would result in the loss of much more.

## A Conservation Crisis

People go to the zoo to see the big and splashy animals: pandas, tigers, elephants, giraffes, and elephants. These animals get all the attention, albeit for good reason. Modern zoos do a great job of pushing the conservation message. Guests learn that many of these charismatic mammals are in trouble. Conservationists love them because they draw in millions of dollars a year to aid in the protection of these animals. But what about the animals without a backbone? Remember the animals that make up around 95% of all animal life? Despite lacking a backbone themselves, these animals serve as the backbone of our ecosystems—and they face more challenges than you might realize. One-fifth of invertebrates face possible extinction in the next 100 years. We are currently living through a mass extinction event, and this one is being caused primarily by humans. And invertebrates such as crayfish are being impacted the most.

Crayfish have been inhabiting the streams, rivers, lakes, and swamps on our continent (and beyond) for as long as it has existed. They have survived at least two mass extinction events, including the K-Pg extinction event, the famous meteor crash that killed the nonavian dinosaurs. But the current extinction event may be far worse for our favorite crustaceans. It's doubtful that they can survive another. Crayfish have been doing crayfish things for a long, long time: creating burrows, serving as food for larger animals, and being little garbage disposals by eating dying plant and animal tissue. But now, half of these species require conservation attention. The top threats are fueled by humans: habitat alteration and, ironically, the introduction of nonnative crayfish species. At least two crayfish species have gone extinct in the last 100 years, with several others being on the brink, inhabiting only a stretch of water no larger than a few football fields.

But before diving into why crayfish populations have declined, it helps to first understand the basics of crayfish biology: what they do, where they live, and eventually, why they are disappearing. Luckily, I am far from the only one interested in these questions. Several scientists have dedicated their lives to these crustaceans. Many of them are just like me, chasing the dopamine rush they experienced as children. Much of what biologists know about crayfish comes from the minds of relatively few individuals who have made huge contributions to our understanding of crayfish.

Sadly, for many species, it is already too late. But many of them can make it back to a healthy state if action is taken. Campaigns to alert the public about these issues exist, but they are often small in scope, and widescale crayfish literacy is still lacking. If the public doesn't realize and appreciate the value of these crustaceans, how would the people in charge of funding their research or conserving them know? I want to change that. I have worked firsthand with many of the scientists on the front lines of crayfish biology, and I hope to tell their stories in hopes of protecting North America's favorite crustaceans.

# CHAPTER 2

# Mystery of the Mudbugs

## Elbow Deep

AFTER LOOKING UP AND DOWN this Pennsylvania farm road for the third time, I get out and suit up. My entire body is still sore from a week of nonstop collecting. Aware of the farmhouses just around the corner, I try to minimize the noise when shutting my car door. I pop the trunk and slip on my waders, caked with a layer of mud on the outside and slick with sweat on the inside. There is signage on the road for an abandoned shooting range and a field of corn stalks across the road. At the base of the cornfield is a roadside ditch that, to me, is perfect. A crayfish biologist in this part of the country has to be a connoisseur of roadside ditches.

Ditches, of course, attract water after rainfall, making them long pits of mud—a perfect habitat for crayfish. How do I know which ditch to explore? The dead giveaway is the dozens of mud piles that look like circular chimneys—a structure permanently burned into my brain after hundreds of hours of searching for them.

I lock the car and hurry away from the shooting range lurking behind me, taking a thirty-second walk up and down the ditch to form my plan of attack. Every few feet, there are haphazard holes in the ground, each with a diameter somewhere between that of a quarter and a half dollar. Half of these holes are hidden underneath structures that look like a toddler's first attempt at pottery.

A typical roadside ditch habitat where you can find burrowing crayfish species. Photo by Zackary A. Graham.

A field littered with burrowing crayfish chimneys. Photo by Shelby Townsend.

Although my presence is entirely legal—I have the proper permits, and Pennsylvania roadside ditches are public property—I always get a weird sense of urgency and anxiety in these scenarios. Maybe it's the stories I've heard of other well-meaning crayfish biologists being chased out of a ditch with gunshots.

I slide my way down into the ditch and pick one of the largest chimneys to work with first, looking for a glisten of wet mud, a telltale sign of recent activity. By gently cupping the base of the chimney with my hands and moving it onto its side, I reveal a single tunnel that's about an inch in diameter. Based on the size of this tunnel and its recent-looking excavation, I know that there is a medium-sized crayfish hunkered down a few feet beneath me. Primarily nocturnal, these animals typically hide out during the hottest parts of the day and only become active an hour or so past sundown.

Using my naked left hand, I peel away mud from the burrow opening to widen the hole until it is twice as wide as when I started. It rained last night, so only about five inches beneath the burrow entrance tunnel, called a portal, is murky brown water. Unlike my now muddy left arm, my entire right arm is protected by a yellow rubber glove that goes up to my elbow. After a final look up and down the road, I clench my fist and wiggle my gloved arm down into the widened tunnel. When my hand slides into the burrow, there is a slurping, suction-like noise, which is music to my ears. It's a tight fit. The goal of this bizarre act is for my fist to act as a plunger and disturb the water in the burrow—and especially the resident crayfish hiding below. Not only is this an odd experience for any human engaging in the plunging process, but I imagine that the crayfish is just as confused. Based on my preliminary probing and the volume of the suction, I estimate that this burrow is just over two feet deep and splits into two tunnels about halfway to the bottom.

Burrow plunging is a methodical process. After shoving my fist in and out of the burrow a half dozen times, I take a short break to stare at the muddy water in the burrow in hopes that the naturally curious crustacean comes to investigate my disturbance. With each bout of plunging, the suction noise gets stronger, which assures me that the water is being properly disturbed. I am now entirely committed to this act—on my hands and knees, covered in mud, repeatedly shoving my arm elbow-deep into the burrow. To me, it's an ideal way to spend my week.

After a few more minutes of laser-focused plunging, I take another short break only to spot a man on the gravel road walking straight at me. I freeze. Despite the late-afternoon heat of July, this man is dressed for winter: weathered boots, grass-stained blue jeans, and a thick lumberjack-looking flannel. By his side is an unleashed German shepherd.

With equal parts embarrassment and fear, I wait for the man and his dog to approach. We awkwardly make eye contact, and he rightfully asks me, "What the hell are you doing!?" I give him my typical spiel, nervously explaining, "I'm a biologist looking for crayfish for my research." But once I mention crayfish, his eyebrows lower in confusion. Like most people, I recognize that this man associates crayfish only with the species that inhabit flowing bodies of water like streams and creeks. He is completely unaware he's been neighbors with burrowing crayfish his whole life. "The types of crayfish I am looking for are called burrowing crayfish, which can live far from permanent water, as they dig down into the soil in search of groundwater," I say, as the man nods his head with a mixture of interest and confusion. Seeing his initial anger diminish, I explain, "Ditches like this are one of the easiest places to collect the crayfish I am after, but I travel all over to find them," as I try to give myself some credibility.

But still sensing his skepticism (he hasn't said a word since his initial exclamation), I then realize that I can give him proof that these animals exist (and that I am not a secret government informant). I give him a step-by-step tutorial of my plunging process and after one final round of plunging and waiting, I detect activity near the surface of the water. Seconds later, a

The Little Brown Mudbug (*Lacunicambarus thomai*). Throughout this species' range, it has an affinity for the wet, nutrient-rich soil on the edges of farms. Photo by Zackary A. Graham.

compact crayfish comes raging out of the burrow with claws flailing, ready to pinch anything in the way. I pin it down and grip the body firmly with my ungloved hand. The man, who has been peering over my shoulder this entire time, lets out a quiet huff. After cleaning off the layer of mud that encased the crayfish, I start a second lecture, in hope that my presence is now justified. By this point, the man's German shepherd is preoccupied with the scent of my car, which is parked off to the side of the farm road.

"This is what I am looking for. This crayfish here is known as the Little Brown Mudbug," *Lacunicambarus thomai*—though I omit the scientific name when talking to him. Gesturing to the burrows, I explain that this burrowing crayfish species is adapted to life in a complex network of tunnels that it excavates and lives in. The crayfish is olive brown, but it has faint highlights of green and red on the tips of its claws, which I point out to the man as he is drawn to the edge of the ditch. And as its name suggests, this crayfish is fond of mud. In further alignment with its name, the crayfish I pulled out of the burrow is relatively little, but in some cases the Little Brown Mudbug can reach almost the length of my palm. I lastly explain, "Aside from crayfish, crawfish, and crawdad, mudbug is another name given to this group of animals. Mudbugs typically refer to the burrowing species." I look down at my mud-covered body. Finally, the man speaks in the stereotypical voice of a hardworking farmer: "I've lived in the area my whole life and never once thought anything of those little mud piles."

This man is like most locals I encounter during my collections: initially confused, partially angry, but eventually welcoming and curious. Now that I've gained his trust, the conversation transforms into a standard discussion about my childhood spent only about a half hour from here, as he becomes friendly and interested. Eventually, he gestures to a nearby field. "The bottom part of my property is swampy and is filled with those mud piles. There's a creek down there too, if you wanna look for your critters," he says. Based on his description, the area could be a potential mother lode. I hurry up the conversation by climbing out of the ditch and placing my crayfish into a compartment of a clear tackle box (which many crayfish biologists call "crayfish hotels") filled with a few other crayfish I collected earlier. He senses my urgency (and excitement at the opportunity to collect crayfish in a new location), and we exchange goodbyes as he continues his evening walk with some newfound knowledge and I go off to find more crayfish before sunset.

A quick wander through the cornfields leads to the edge of a field on his property. Even with his permission, it feels strange tromping around someone's yard in plain view of their front porch, not to mention the gaudy rooster sounding his alarm just a dozen yards away. With each step through the patchwork of mud and grass, my wader boots sink halfway up to my ankles. It doesn't take long to find a large patch of moist, muddy bottomlands next to a stream. Scattered throughout this private patch of field are dozens of burrows spread as far as I can see. Exactly what I was hoping for. I plunge until the daylight fades, proudly brandishing my crayfish hotel filled with a dozen or so mudbugs.

## Describing Diversity

After Horton H. Hobbs Jr.'s initial encounter with crayfish during his freshman year of college, the remainder of his life was filled with everything and anything crayfish (an experience I can relate to). Wherever Hobbs went, he brought his collection of jars with him, each containing anywhere from one to thirty crayfish and filled with ethanol, the ultimate preservative liquid for dead crustaceans. Accompanying each jar was a label with the exact date, location, and scientific name of the species within. Hobbs's collection bled from his work to his home, filling up his basement and garage. He amassed hundreds of boxes filled with thousands of glass jars, all containing preserved records of the crayfish he collected.

This sixty-year collection ended up with Hobbs at the Smithsonian National Museum of Natural History, in Washington DC, where it still resides today. Hobbs's collecting led him to travel throughout North America (and the world) with the goal of documenting and describing the diversity of crayfish that he found. His collection serves as a physical record of his findings and will be used by biologists for centuries to come. Since Hobbs is the closest thing there is to a rock star in crayfish biology, every time I visit his collection it comes with an awe-inspiring feeling—rows on rows of jars stacked across ceiling-high shelves that showcase his life's work. Not only does this collection hold a part of Hobbs's legacy, but it also represents the legacy of crayfishes and their ability to dominate freshwater habitats throughout eastern North America, from the widest rivers to the muddiest ditches.

HOBBS KNEW THAT some crayfish inhabit streams, lakes, and rivers but others construct complex burrow networks that can be found far from any permanent body of water. Complicating matters even more, Hobbs realized that some crayfish are flexible: They are capable of inhabiting both permanent bodies of water and burrows. But without proper definitions and descriptions of these lifestyles, he had trouble in his quest to describe the diversity of crayfish. Hobbs invented his own system, which classified crayfish into one of three groups based on their ability to construct burrows and their reliance on these burrows. These classifications are best described in Hobbs's seminal 1981 text, *The Crayfishes of Georgia*, which provides a 549-page detailed account of general crayfish biology, the three broad groups of crayfish, and a survey of every known crayfish in the state of Georgia at the time. (As a fun aside, Hobbs's wife was named Georgia, and one of his favorite non-crayfish pastimes was to play piano duets with her after evening dinners.)

Hobbs's classification system allows biologists to understand and appreciate the diversity of size, shape, and behavior found in crayfish populations. The first of Hobbs's three classification groups contains those crayfish that the average American is most familiar with, which Hobbs called tertiary burrowing species. These crayfish are generally large bodied, and they carry around their bulky claws in a rather awkward manner (with a few outliers, just like everything in the complicated world of biology). Because of their typically large size, tertiary burrowers are highly adapted to life in flowing water; they create only small, rudimentary burrows underneath rocks or other large substrates in permanent bodies of water. Tertiary burrowers simply don't have the correct tools to excavate complex burrows. It would be like trying to dig a tunnel with a sledgehammer.

A Bottlebrush Crayfish (*Barbicambarus cornutus*). Photo by Guenter Schuster.

Tertiary burrowing crayfish rely on natural shelters to do all the work; they just wedge themselves underneath. These shelters provide refuge from predators during the day and are vigorously guarded during the mating season. Without a rock or fallen tree trunk to hide under, a crayfish is just a large chunk of protein waiting to be gobbled up. And because they share the aquatic environment with large fish predators, tertiary burrowers often have large claws, which they can use to ward off any intruders from their burrow.

The crayfish that best embody the tertiary burrowing species are the Bottlebrush Crayfish in the genus *Barbicambarus*—which are named for their bearded, fuzzy antennae that resemble bottle brushes. *Barbicambarus* are the largest crayfish in the United States. Every year or so when I feel burned out, I take a trip to visit these massive crayfish, which is like a reset button for my mental health. The first time I collected one of these behemoths in a Kentucky stream, a middle-aged couple shouted at me while I was in the middle of a fifty-foot-wide stream. With rushing water drowning out whatever they said, I assumed they were wondering what I was doing, so I held up a *Barbicambarus cornutus* nearly the length of my forearm and the couple shouted in amazement (rightfully so—*Barbicambarus* are awesome). After their excitement faded, I continued on with my business, but the couple's newfound awareness of foot-long crayfishes must have acted as an instant aphrodisiac, because they (strangely) proceeded to make out on the riverbank while I continued to search for these animals.

Because *Barbicambarus* are too bulky to create complex burrows, they thrive in the cavities they create underneath large shelters. If you want to find *Barbicambarus*, you need to lift slab rocks the size of your dinner table. Slab rocks are large but relatively thin, which makes them ideal for flipping. And when it comes to searching for *Barbicambarus* or other tertiary burrowers, lifting the largest slab rocks always yields the largest crayfish—which leads to lower-back injuries being common among crayfish biologists.

TERTIARY BURROWING CRAYFISH inhabit bodies of water that are permanently flowing year-round, but not all water is permanent. During drought or because of the naturally fluctuating water levels of some water bodies, a once consistent flow of water may turn into a dry creek bed. The crayfish that live in these unpredictable environments can be flexible in their burrowing

Typical burrows created by secondary burrowing crayfish on the edge of a stream bank. Photo by Zackary A. Graham.

ability; Hobbs called them secondary burrowers. Secondary burrowers can create simple burrows under rocks when water is present (just like tertiary burrowers), but when the water level falls, they are also able to excavate simple burrows to reach the receding groundwater.

Growing up, the crayfish that I caught the most was a secondary burrower, the Rock Crayfish (*Cambarus carinirostris*). This is one of hundreds of crayfish species that live under rocks, which demonstrates the pitfalls of using common names instead of scientific names. During the summer, when water is flowing, Rock Crayfish are abundant under rocks in streams. But during the winter, when the water recedes, these crayfish retreat into burrows they dig under rocks and in the edges of the stream bank. I never realized it at the time, but the edges of my childhood streams are littered with crayfish-sized holes that run diagonally to the underground water. These holes are the emergency tunnels that Rock Crayfish create to access groundwater when the normally flowing stream water disappears.

To see the differences between tertiary and secondary burrowers, compare their bodies, because the body of a secondary burrowing species differs from that of the hefty tertiary burrowing species. Secondary burrowers are generally small to medium in size, and their claws are less exaggerated. These crayfish also usually lack spines, which tertiary burrowers often possess. Smaller, smoother bodies help secondary burrowers navigate their burrows. But even though secondary burrowers can create impressive tunnels to reach groundwater, they are no match for the labyrinths created by the true burrowing crayfish, or what Hobbs called the primary burrowers.

AS YOU SAW IN MY ROADSIDE ditch adventure at the start of this chapter, primary burrowing crayfish take things to a different level. These animals rely on their burrow for nearly every aspect of their life. Instead of relying on water from streams, lakes, or rivers, they rely entirely on the groundwater hidden beneath our feet. In a roadside ditch or near a stream, the water may only be a few inches below the surface and the burrows in these areas may not be overly complex. But other primary burrowers may live far from water, and their burrows can be over ten feet deep, with numerous tunnels, chambers, and entrances. Despite the general public's lack of knowledge that these animals exist, there are well over 100 crayfish species that live this elusive lifestyle.

Primary burrowers are perfectly adapted to excavate their burrows and access the groundwater below. Their bodies have been modified to life within the confines of their burrow. Primary burrowing crayfish bodies are tall rather than wide. Because life underground involves lower oxygen levels

A secondary burrowing crayfish species, the Red Swamp Crayfish (*Procambarus clarkii*), about to emerge from its burrow with a fresh pile of mud. Wikimedia Commons.

and decreased airflow, this taller body gives them plenty of room for their gills, which all crayfish use to breathe. Further, the claws of burrowing species are generally small and robust, which allows the crayfish to tuck their claws in close to their body so they can crawl around their burrow. At the other end of their body, the "tail" of primary burrowing crayfish, which is the abdomen, is also smaller than the strong muscular tail of a tertiary burrowing species. In tertiary burrowers, a large, muscular abdomen enables a strong "tail-flip" escape response, in which the muscles are contracted and the crayfish can dart away from a predator. Because primary burrowers rarely encounter predators in open water, a large abdomen is unnecessary. Together, primary burrowers and secondary burrowers are typically referred to as "burrowing crayfishes" or "burrowers," whereas tertiary burrowers are referred to as "stream-dwelling crayfishes" or "stream dwellers."

One trip at a time, building from the bottom to the top, burrowing crayfishes gather mud and place them into organized pellets outside their burrow. Sometimes burrowers will craft these mud pellets into a perfectly formed chimney, and other times they will just place the mud immediately outside the burrow without any attempt at chimney construction. Although the exact reason that crayfish construct these chimneys is debated, recent evidence suggests that the chimneys are ventilation structures that help

# *DISTOCAMBARUS,* A GROUP OF EXTREME BURROWERS

Of all of the primary burrowing crayfishes in the United States, those in the genus *Distocambarus* may be the most extreme. This group encompasses four species that are endemic to the foothills of South Carolina and one species whose range creeps across the Savannah River into Georgia. *Distocambarus* can live a significant distance from a permanent body of water, which has led many biologists to classify them as terrestrial, based on their ability to successfully live away from water. Some of these species, such as the Piedmont Prairie Burrowing Crayfish (*D. crockeri*), have long, gangly alien-like arms that seem like they would be a hindrance to burrowing. Despite this, *Distocambarus* burrows can reach well over six feet, and many scientists believe these crayfish can live more than a dozen years.

A Piedmont Prairie Burrowing Crayfish (*Distocambarus crockeri*).
Photo by Zackary A. Graham.

pull in a breeze and aerate the burrow. Without this aeration, oxygen levels become so low in a burrow that neither the water nor the air allows the crayfish to breath efficiently. Even so, crayfish burrows often lack a chimney, and crayfish seem to survive perfectly fine even if a chimney is "capped," or closed off.

Hobbs admired the primary burrowers and knew there was much to learn about these animals. Although Hobbs provided biologists with a framework to understand the different burrowing strategies, catching burrowers is still difficult, and it takes a lot of time and energy. Early on in my career, my average rate of burrowing crayfish collection was around one crayfish per hour on a good day. Now with more experience, these numbers have doubled or tripled. Still, collecting these animals is time consuming, and because of this, the biology of burrowing crayfishes remains something of a mystery, with a lot of our understanding coming from anecdotes. Such one-off observations are shared in the scientific literature, through emails, and at conferences. But full-scale studies on these animals are rare. There is a lot to learn about the mystery of the mudbugs.

## The Secret Lives of Burrowing Crayfish

As I make my way through the mud, I'm surrounded by the stench of Skunk Cabbage (*Symplocarpus foetidus*). For me, this smell brings more nostalgia than nausea. When the cabbage-like leaves of this plant are bruised, it releases a fragrance akin to that of decaying flesh. And unfortunately for me, Skunk Cabbage grows in large patches throughout muddy bottomland habitats—ideal locations to search for burrowing crayfish.

This is the second day in a row that I have been plowing through Skunk Cabbage patches and plunging crayfish burrows—with only two crayfish to show for it. The crayfish that I am looking for is one of my favorites. I am targeting the appropriately named Blue Crawfish (*Cambarus monongalensis*). The Blue Crawfish is a mountain specialist that loves to excavate burrows near Skunk Cabbage patches throughout Pennsylvania and West Virginia. The species name, *monongalensis*, combines *monongal*, from the Monongahela River, which surrounds this species' distribution, and the Latin suffix *-ensis*, which means "originating in." This is a stout species, reaching only a few inches in length. And because it has a deep-blue body and contrasting splashes of orange on the tips of its claws, many people believe this animal is fake when I show them pictures on my phone.

This species was named in 1905 by Arnold E. Ortmann, who traveled by railroad and on foot over 11,000 miles (he counted) throughout Pennsylvania

A Blue Crawfish (*Cambarus monongalensis*). Photo by Zackary A. Graham.

to document crayfish. During Ortmann's time, the term "crayfish" had yet to be solidified by the scientific community, and therefore many species he named have "crawfish" in their common name. Ortmann collected these blue crustaceans from burrows ten minutes outside of my hometown and five minutes from the Carnegie Museum of Natural History in Pittsburgh, where he was the museum's invertebrates curator. At the time that Ortmann named this species, the Blue Crawfish was found "everywhere" throughout Pittsburgh, according to his work. But now only a few disjunct populations remain in Pittsburgh's public parks. Because burrowing crayfish require specific habitats with high water tables and soil amenable to burrowing, they will often form a large congregation of burrows, called a colony. Colonies are roughly correlated to populations, which are defined as a group of organisms that interact and can mate with one another, but crayfish biologists use the term "colony" to informally describe a cluster of burrows in one habitat.

Back in the Skunk Cabbage patch, I locate a dozen burrows within my first few steps. Each burrow is tucked underneath the overhanging leaves of this plant, which, like the crayfish I am after, prefers to live in moist and muddy plots of land. While tallying burrows, I notice the looping tracks of a raccoon embedded in the mud—likely made the night before, in search of a blue-colored treat.

After aimlessly following the raccoon tracks, I pick a good-looking burrow and then get down to business to start my plunging routine. After only a few plunges, I can feel that the suction isn't quite as strong as normal. As I go deeper into this burrow, the walls of the tunnel suddenly disappear, and my clenched fist can freely open and wiggle around. With my cheek resting on the mud and my entire arm underground, I am swashing around in a soccer-ball-sized resting chamber filled with water. In my experience, only adult crayfish construct these massive resting chambers, which suggests that a mature Blue Crawfish is just out of my reach.

With a better sense of the burrow's architecture, I continue with another round of plunging and waiting. Typically, the resident crayfish will creep up from its resting chamber and flick its antennae around the top of the water as an initial safety check. But this time, instead of the antennae of an adult, I see a small army of baby crayfish crawl out the burrow. Each of them is no larger than a grain of rice as they awkwardly meander around the burrow's entrance. This is a true *National Geographic*–type moment for me, as I hear David Attenborough's narration in the back of my head.

Unlike their vibrant parents, these craylings (the scientific term for baby crayfish) are translucent blue (sometimes almost white) and lack the orange highlights. One by one, eighteen craylings emerge. And then as if she is curious about where her babies went, a large blue female crayfish finally shows herself. After feeling guilty for breaking up their underground play session, I let the mother and her craylings wander back into their burrow.

For decades, crayfish biologists have been digging up maternal crayfish living with their offspring. And each time they make such an observation, speculations abound that burrowing crayfish might be social creatures. But because these animals are nocturnal, spend most of their lives underground, and are a pain in the rear to collect, these speculations used to be just that— entirely speculative. But now, our understanding of the social and parental nature of these animals is starting to come together.

JAMES NORROCKY was a lucky man. A colony of Digger Crayfish (*Creaserinus fodiens*) lived in the yard of his Ohio home. After a hard day of delivering mail, Norrocky would start his night shift as a crayfish biologist—digging burrows in his yard and the surrounding ditches. Norrocky never received any formal education as a scientist. But he was taken under the wing of a local astacologist, Raymond Jezerinac, who was known for taking motivated individuals and teaching them the ins and outs of crayfish biology, regardless of their credentials.

Digger Crayfish are tricky because they love the short-lived puddles that form in roadside ditches after a heavy rain. But when these ditches dry up, they construct burrows to reach the underground water table. This flexibility makes some biologists classify them a secondary burrowing species, but because they do not rely on permanent water, many consider them to be primary burrowers. And Digger Crayfish seem to thrive in the habitat right between the road and people's lawns, which can become a nuisance for non-crayfish-inclined homeowners. I have seen lawn mowers plow over one-foot-tall crayfish chimneys before, with the person doing the mowing completely oblivious to the colony of crayfish living underneath their yard.

With a yard full of burrows at his disposal, Norrocky wanted to learn about the biology of these understudied animals: how they mated, how long they lived, and whether they were moving throughout the colony. But the soil wasn't always amenable for plunging in this area, so he resorted to full-on digging, which is exhausting, especially when the crayfish burrows can be several feet deep.

A Digger Crayfish (*Creaserinus fodiens*). Photo by Zackary A. Graham.

After getting fed up with the time-consuming burrow-digging process, Norrocky decided to make a custom burrowing crayfish trap. If this trap worked, not only would it put his crayfish collection process on autopilot, but it would also resolve his wife's and neighbors' complaints about the scattered holes dug throughout the yard.

And voilà: Norrocky created the first-ever burrowing crayfish trap. The Norrocky Burrowing Crayfish Trap, as it is now known, is simple. The trap consists of a PVC pipe that gets shoved at a slight angle into the entrance of a crayfish burrow. If the burrow entrance was topped with a chimney, Norrocky would remove the chimney to get a clear view of the entrance. Within the pipe is a small one-way metal flap; once the flap is lifted, the crayfish cannot pass through the flap again. So if a crayfish wanders out of its burrow and goes past the metal flap, it is unable to reenter the pipe and get back down.

Norrocky went on to use his new trapping invention over the next seven years. Each time he would trap a crayfish, he would record its sex, take standard body size measurements, and give it a small marking on its abdomen— a unique identifier for each crayfish—before placing it back in its burrow. By marking and releasing these animals, he was able to determine how long each crayfish stayed in its burrow and where/whether it was moving about in the colony. This technique, called mark-recapture, is a staple in biology, because it allows biologists to understand the marked animals' fine-scale movement patterns and behaviors. Norrocky was not the first to conduct a mark-recapture study on crayfish, but he is one of the only people who has attempted to do this on a burrowing crayfish species. Over the course of the seven years that he checked his traps and marked and recaptured these crayfish, Norrocky collected 496 Digger Crayfish. Based on my average collecting rate of one burrowing crayfish per hour that I was used to, I am jealous of that number, because no exhausting digging or plunging was required.

One of the reasons that burrowing crayfish are so hard to collect is that they were historically thought to rarely leave their burrow. Scattered observations of surface activity existed, but they were infrequent. On his collecting trips, Hobbs often found burrowing crayfishes crossing roads at night when it was raining and thought these animals only left their burrow to forage for food or find a mate.

Norrocky had witnessed nighttime excursions like those Hobbs had reported, so he had a hunch that these nighttime movements were common. But the extent of the movement and activity was unknown. To Norrocky's surprise, he found that these crayfish were serial burrow hoppers—crawling from burrow to burrow throughout his yard and beyond. Over 100 crayfish changed their burrows during his study period. Some of their movements were quite the trek—up to 150 feet away from the initial burrow in which they were captured. For a two-inch crayfish, this distance is significant. And because Norrocky placed traps over only a limited number of burrows every night, such excursions were likely much more common than what he could record.

Sometimes these movements landed the crayfish in an empty burrow. But other times, crayfish trekked across the yard and stayed in a burrow that was presumably occupied by another Digger Crayfish. And strangely enough, in one instance, Norrocky found a completely different species, the Great Plains Mudbug (*Lacunicambarus nebrascensis*), sharing a burrow with a Digger Crayfish. These were adult crayfish sharing the tight quarters of their burrow with members of the same species—and in one case, a completely different species! For a crustacean, this type of sociality is the exception, not the rule. Take the crayfish's closest living cousins, clawed lobsters, like the ones you see in the grocery store or the local seafood joint. These lobsters, such as the American Lobster (*Homarus americanus*), have their claws banded together when forced to live in these communal tanks. But in nature, without their claws restrained, the lobsters would never tolerate being in such close proximity to one another and would instantly break out into a territorial dispute.

Norrocky recorded thirty instances of burrow sharing between adult crayfish, and these pairs were observed in every combination of age and sex: Adult males shared burrows with adult males; adult females shared burrows with adult females; adult males shared burrows with juvenile females; and so on. In one instance, the same two crayfish were repeatedly caught in the same burrow for over three and a half years! For animals that were once thought to be solitary, these findings have made crayfish biologists reconsider the social lives of burrowing crayfishes.

In my opinion, this part-time work by James Norrocky is one of the single greatest research efforts in crayfish biology. Before this work, astacologists knew that burrowing crayfish could sometimes be found in the same burrow as another crayfish, but there was never much credit given to these observations. Such burrow-sharing reports were few and far between, leaving scientists with nothing but speculation to work with. But now, with seven years of data thanks to Norrocky's work, there is strong support for the social nature of these reclusive animals, which brings up even more questions. Why are these animals so friendly with one another? And what exactly is going on in these burrows?

EVEN FOR A BURROW-DIGGING specialist like the crayfish, excavating a burrow is an arduous task. And the colonies that crayfish form can occur only on small patches of land where the conditions are just right—it can't be too dry, or too sunny, or too far from the groundwater. These requirements leave burrowing crayfish unable to spread out and explore greener (or more realistically, browner) pastures. Being confined to one area is thought to have forced these crustaceans to develop strong social skills and care for their young. The relationship between burrowing and sociality has been raised for other burrowing species as well, such as ants and naked mole rats, which form amazingly complex social hierarchies. Without one another, these animals are unable to survive.

The semicommunal nature of burrowing crayfishes is completely different from that of the stream-living species. In the stream, competition is fierce. Fighting is a way of life; stream-dwelling crayfishes do not tolerate each other one bit. By contrast, in a colony of burrowers, the crayfishes' behavior around each other is much different, and this is especially true for their relationships with their young. The passive nature of burrowers is apparent when you interact with them. Place ten stream dwellers in a bucket, and you will witness plenty of aggression. Place ten burrowers into a bucket, and you will generally observe their passive demeanor with one another. However, some burrowers can pack a punch with their claws and are not afraid to unleash them on an unsuspecting crayfish or human nearby—but I have found that this is generally the exception and not the rule. Overall, sociality and a lower level of aggression is unique to the burrowers.

A benign bucketful of burrowers begs the question: What exactly goes on underground in these shared burrows? Is there some kind of strength-in-numbers game going on? Plenty of other animals that live underground are known to team up to thrive in their environment. But why do many burrowing crayfish tolerate each other? What's the benefit?

Long story short, crayfish biologists don't know. When two adults share a burrow, the possibility for mating is likely—unfortunately, this hypothesis fails to explain the commonality of same-sex burrow sharers. Maybe the crayfish completely avoid each other while in the burrow. Maybe they share food that has been stored in the burrow. At this point in time, there is only speculation in this realm of crayfish biology, but crayfish enthusiasts like me (and maybe you, after reading this book) are champing at the bit to figure it out.

Another puzzle in the world of the burrowing crayfishes relates to their parental care. At the very least, parents let their offspring stick around for a while, until the young crayfish can forage for themselves and excavate their own burrow. This is like the all-too-familiar scenario where a newly gradu-ated college student moves back in with their parents. In theory, the recent grads can go out into the world by themselves, but they will struggle early on—so living with parents serves as a safety net in a time of uncertainty. This greatly contrasts with the parental care of stream-dwelling species, which kick out their young the second they can fend for themselves. And if the juveniles of stream-dwelling species stay around for too long, they will become their mother's lunch. Mother-offspring cannibalism is uncommon in burrowing species.

Although this leaves more questions than answers, sociality is clearly a trait that allows burrowing crayfish to thrive in a unique environment. It is my hope that future research (by both scientists and nonscientists) will help us understand the nature of these animals within my lifetime. For example, burrow sharing between parents and their offspring appears to be a fairly common trait among all burrowing crayfish. However, what is less clear is whether parents provide their offspring with any direct resources in their underground shelter or whether they simply put up with them and allow them to hang around. A consistent theme in this chapter is that the naturally reclusive nature of burrowers and the difficulty of their capture makes them a huge mystery. This fact is both discouraging and exciting, and it will surely take some creative minds to figure out these mysteries.

As far as burrowing crayfish go, perhaps the biggest question that re-mains unanswered pertains to a part of their biology that I've glossed over thus far: their color. Hidden underground in these mud burrows, many of the resident crayfish are puzzlingly colorful—and biologists continue to debate why.

## The Million-Dollar Color Question

I previously mentioned one of the gems hidden beneath the mud around my hometown of Pittsburgh, the Blue Crawfish. Without going into too much detail, I had hoped that reading about the existence of a bright blue crayfish with orange highlights made many of you pause and think for a second. Why would a nocturnal crayfish that spends its life hidden away in a subterranean maze of mud and muck wear a color that could rival the brightest jewel? It's as if nature decided to hide a gem deep underground and then left us to ponder its purpose.

Some things in nature just don't make sense, and it drives biologists crazy. In crayfish biology, nothing is more poorly understood and hotly debated than the conspicuous colors (blues, reds, oranges, and purples) of burrowing crayfish. You would think that an animal that spends nearly its entire life hidden underground would be drab, which is true for many burrowing crayfish, such as the Little Brown Mudbug and the Digger Crayfish. Dull colors like browns, tans, and grays make plenty of sense for an animal that lives underground in the mud. Without any light in a burrow or interaction with the light at the surface, you would expect bright colors to be forgone. Just glance at some of the most diverse underground-living arthropods in the world, the ground beetles, which are commonly dark and earth toned, compared to their surface-living counterparts, which may be glistening gold or green. Furthermore, even if one of these burrowing animals does emerge from a burrow, it makes sense for it to be camouflaged in the surrounding soil and mud.

Despite the intuitive coloring of camouflaged burrowing animals, burrowing crayfish in North America (and beyond) are often brightly colored—approximately 25% of them exhibit blue, orange, red, purple, or some combination of these colors, nearly the full ROYGBIV spectrum. Furthermore, these base colors are often accentuated by highlights or bands of color throughout their bodies. Even more confusing is that when the crayfish are visible to the outside world, they may be caked in mud, which conceals their colors altogether. I have spent (and will continue to spend) countless nights lying awake puzzling over these colorful burrowers—trying to make sense of it all.

TO START TO UNDERSTAND crayfish coloration, a quick lesson on the biology of color and what exactly it means to be colorful will help. Color is our perception of specific wavelengths of visible light that is reflected off an object. So whenever a human views a red stop sign, the sign is reflecting wavelengths of light that our eyes perceive as being red; all the other wavelengths that humans perceive (blue, green, etc.) are absorbed by the sign. But all of this depends on the visual capabilities of the receiver. Humans have above-average color vision, with three different color receptors (cones), each of which is specialized to process a specific wavelength. The three color-related cones give humans trichromatic vision—the ability to process three color channels.

If an animal is colorful, the color comes from one of two different pathways. Some colors come from the consumption of color pigments, which alter the animal's appearance (often called physiological color). The red feathers of a male cardinal are the textbook example of physiological color. The reddest cardinals eat the most pigment-containing foods, such as berries and other fruits. When deprived of these pigment-rich foods, they lack their famous red color. Physiological color pigments are responsible for most of the yellows, oranges, and reds expressed by colored organisms. In the second pathway, color comes from the underlying structure of the materials that reflect the color (often called structural color). In nature, structural colors are often associated with blacks and blues, although blue is among the rarest colors in nature. Few animals have evolved the unique innovation of blue coloration, with notable exceptions in the massive wings of the blue morpho butterfly or the feathers of a blue jay.

Just as birds have feathers that express color, crayfish have color in their exoskeleton, the hardened outer layer of defense in arthropods. Several individual sublayers overlap to create the exoskeleton. Each sublayer has a different thickness and different properties. And throughout this layering of armor are pigments, which are held in different densities and locations. With this unique overlapping network of pigment layers, crayfish express physiological color that results in a spectrum of colors, from a river-rock gray to pure blue. Because these colors come from external pigments, this also means that the diet of the crayfish—and its ability to consume pigment-rich foods—can alter its color.

Take a look at one of Kentucky's and Tennessee's most prized possessions, the Valley Flame Crayfish (*Cambarus deweesae*). This primary burrowing crayfish can be bright red and orange—which explains its common name. But across its range, you may find a half dozen or so crayfish with different color variations, called color morphs. *Cambarus deweesae* color morphs can be reddish orange, solid blue, blue and orange, blue and red, and even a bland pewter gray. Each of these color morphs results from different placements and densities of pigments within the multiple layers of the animal's exoskeleton. Data suggests that the color morphs may be considered distinct species in the future, but their variation is impressive nonetheless.

In the orange and red morphs of *C. deweesae*, there are likely high levels of the carotenoid pigment astaxanthin, which reflects red wavelengths of light and absorbs the remaining wavelengths. When you boil a crustacean (crayfish, lobster, shrimp, etc.) and its exoskeleton turns bright red, this is because all other pigments get destroyed during boiling—but the red-reflecting astaxanthin pigment remains.

Although blue is most often a structural color produced in nature, crustaceans produce blue through pigment-based biological color thanks to a crustacean-specific pigment called crustacyanin. Thus, the blue regions in some *C. deweesae* morphs or the blue color in *C. monongalensis* are likely chock-full of crustacyanin, with little to no astaxanthin coming through.

The blue-and-red morph of the Valley Flame Crayfish (*Cambarus deweesae*).
Photo by Guenter Schuster.

To confirm these ideas, I have boiled the claws of a *C. monongalensis* before, which after only a few minutes in boiling water resulted in a burnt-orange claw, because the crustacyanin degraded from the heat. This blue-turned-orange claw now sits in a small vial proudly displayed on my desk.

In the multicolored red-and-blue *C. deweesae* morphs, both astaxanthin and crustacyanin are present, but in different locations in the exoskeleton and in different densities. In a more generic-colored crayfish such as the pewter-gray color morph of *C. deweesae*, there are likely several layers mixing to create a bluish-gray color. Anyone who has haphazardly mixed paint colors is aware of the resulting shades of browns and grays.

WITH AN UNDERSTANDING of *how* crayfish get their colors, we can start to answer the million-dollar color question of *why* some crayfish are cryptic whereas others are colorful. The way that animals look and behave usually has some sort of benefit to the animal, because if a trait has a negative effect, then the individual will often not be able to survive or reproduce. But if a trait is beneficial, then it will be passed on to the next generation—a classic case of evolution through natural selection. So what purpose do crayfish colors serve? What is the benefit? And is there a cost?

Camouflage is a good place to start, because it is easy to imagine how an animal that can successfully blend into its background may benefit from not being detected by predators. This is clearly the case for the stream dwellers, because the browns, tans, and grays all help the crayfish blend in with rocks, leaves, and twigs, which reduces their chance of becoming lunch. Stream-dwelling crayfish also exhibit patterning, with dots, splotches, and stripes that presumably aid them by further minimizing detection by predators. For some of the burrowing crayfish, camouflage can explain their colors. Both the Little Brown Mudbug and the Digger Crayfish are examples of burrowing crayfish species with inconspicuous, camouflaging colors. Though these animals rarely leave their burrows, when they do, it makes sense for them to blend into the environment and not stick out like a sore thumb.

Strangely, sticking out like a sore thumb is exactly what many burrow-ers do. A blue or red crayfish on the forest floor is a walking target that is asking to get eaten. But since these crayfish are primarily nocturnal, does it even matter? Maybe their nighttime activity means that they can keep their bright colors and avoid being detected—and thus avoid not getting picked off by their top (also nocturnal) predators, like raccoons and owls. Indeed, at night, whatever color vision crayfishes' predators do have will be dulled. So at night, being brightly colored may not be a bad thing. Although their coloration might not be a detriment to the crayfish at night, it still does not provide a clear benefit.

One theory is that bright burrowing crayfish colors are the opposite of camouflage. Maybe the crayfish are brightly colored so they can be spotted —like a warning signal. This type of antipredator warning coloration, called aposematism, tells predators "I am dangerous, don't mess with me." Warn-ing colors are often associated with toxins, like in the poisonous monarch butterfly and the venomous coral snake, to name a few. But there is no evi-dence that crayfish are toxic. Beyond signaling poison or venom, bright colors might serve as a warning that the crayfish is a threat, cautioning po-tential predators that a powerful pinch could await them.

Although quite passive with each other, burrowing crayfish are quick to strike when they are threatened by a predator—especially the humans that attempt to catch them. Because they live in the confines of snug burrows, their claws are usually compact and strongly built. Some burrowers, like one of the most badass-named crayfish, the Texas Prairie Crayfish (*Fallicambarus devastator*), have a massive, hardened bump on their claws that can puncture human nails—truly devastating. Similar damage could be dealt to any of the dozen natural predators of crayfish, so perhaps coloration is a way to warn them off. Although the antipredator role of the crayfish color makes intuitive sense, there is no strong support for this theory.

Adding to this, there is no evidence that conspicuous colors are used in mating. Both male and female crayfish express the same colors. While other species do use coloring to attract suitable partners—think of the brilliantly plumed male peacock, who must dazzle to be selected by the aesthetically plain peahen—male and female crayfish boast similar coloring, making mate selection an unlikely theory to explain the wild colors of burrowing crayfish.

DR. GUENTER SCHUSTER has been catching, describing, and photographing crayfish for over thirty years. Without him, the photos in this book would not exist, because not only did he take many of them (thank you, Guenter), but he set a new standard for crayfish color photography. Like others before him, Schuster was amazed by the colors of burrowing crayfishes, especially throughout Kentucky, one of the best states for an astacologist to live in. During his years working in one of the world's hot spots for crayfish diversity, not only did Schuster take notes on color, but he also focused on the patterns of color across species. As Schuster highlights, though two crayfish species may be similarly colored, there may be vast differences in the color patterns of these animals. And Schuster thinks that color patterns are not only useful for humans' ability to distinguish between species, but they also may allow crayfishes to distinguish between each other themselves. I watched in awe as Schuster presented these ideas as the keynote speaker at the twenty-second meeting of the International Association of Astacology (yes, there are scientific meetings dedicated to crayfish biology). His calm, steady, and experienced voice guided the audience through a stunning showcase of photographs and insights, revealing the diversity of crayfish colors and color patterns. Witnessing this presentation was a revelation for me, as I realized I wanted to bring crayfish into the spotlight for as many people as possible, just as Schuster had done with his photographs and ideas about crayfish colors.

As Schuster explains, color patterns are the ways that color is expressed throughout a crayfish's body. These color patterns result in spots, stripes, bands, and saddles, and the patterns may exist to aid in differentiating species. Some stream crayfishes, such as the Freckled Crayfish (*Cambarus maculatus*), are littered with freckles of coloration on their entire exoskeleton.

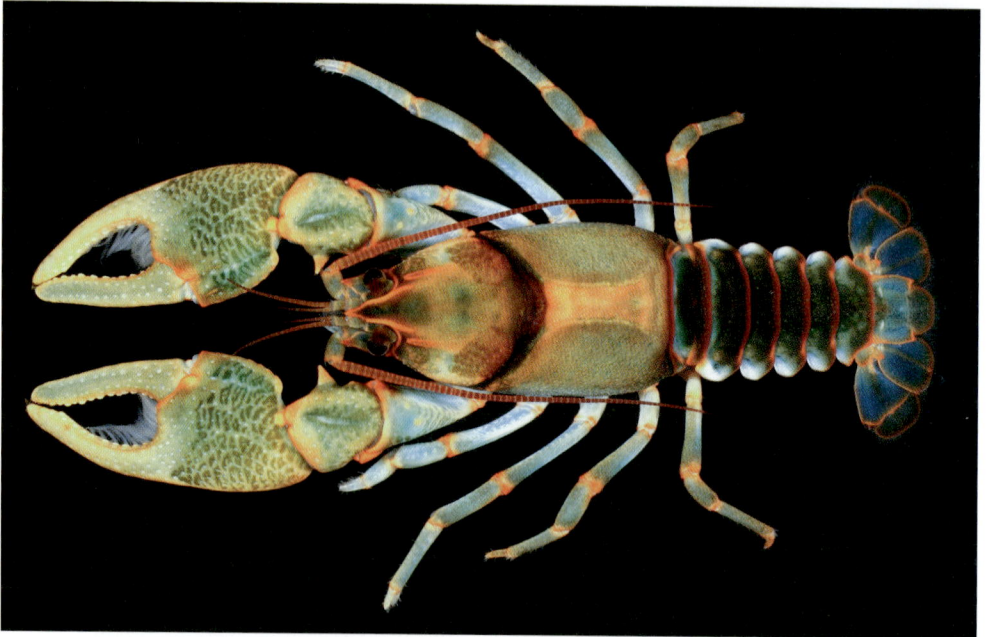

Examples of variation in the color patterns exhibited by crayfish: *Top:* The Freckled Crayfish (*Cambarus maculatus*). Photo by Chris Lukhaup. *Bottom:* The Greensaddle Crayfish (*Cambarus manningi*). Photo by Guenter Schuster.

Others, such as the Greensaddle Crayfish (*Cambarus manningi*), have distinct banded lines and contrasting patterns on their bodies.

One line of evidence that supports Schuster's species recognition hypothesis is that in many species (including burrowing species), the tips of their claws often have a brightly colored band. Our blue friend, the Blue Crawfish, provides a perfect example of this contrasting claw pattern: Its primarily blue claws are highlighted with orange bands at the tips of their pinching fingers. Wielding claws with contrasting colors, burrowers often sit at the entrance of their burrow with nothing outside except their claws, which may serve as a signal of "This burrow is occupied." Remembering the color wheel from elementary school art class, blue and orange are contrasting colors, which illustrates the potential for this color pattern to be used for some sort of species recognition or detection.

With all these theories of crayfish coloration (camouflage, predator defense, mating, species recognition), it is always assumed that there is some benefit to being colorful. Thinking this way can be counterintuitive, because not every trait has to be beneficial. In fact, most cave crayfishes lack pigmentation altogether and appear translucent white. If the color does not nega-

A Southern Cave Crayfish (*Orconectes australis*). Like many cave species, this one lacks pigmentation altogether. Photo by John Abbott/Abbott Nature Photography.

tively influence the crayfish's ability to survive and reproduce, the colors can stick around—they may be evolutionarily neutral. According to this neutral theory, color may be the by-product of some other biological process or the remnant of a historically beneficial trait that is no longer useful. The human belly button is a prime example. Belly buttons are not beneficial or harmful; they are just a by-product of the necessity of the umbilical cord.

Could the bright coloration of crayfishes be neutral—neither helping nor hurting? A recent study that my collaborator (and good friend) Dylan Padilla Perez and I published supports this controversial theory. After analyzing hundreds of crayfish species, we found that bright colors like blues, reds, and oranges have evolved more than fifty times in crayfishes and that these colors were more likely to evolve in primary burrowing crayfish. We also noticed that mutations that influence crayfish color are common, which presumably make them unable to express some of the pigments discussed earlier. In theory, if such mutations occurred in burrowing crayfish populations and they just stuck around, the colors may not be beneficial and they may not be harmful, but they could be neutral, serving as an example of a "happy evolutionary accident." And since burrowing crayfish are commonly isolated, with little ability to mix their genes with other crayfish populations, the colors may just stick around. The use of genetic technologies will likely give us an answer regarding the "neutral theory of crayfish color" in the future.

The reality is that biologists don't know the answer to the million-dollar color question. Understanding the role of coloration is complex, with dozens of unanswered questions. Likely, several of the theories play a role. These split opinions and the conflicting evidence regarding crayfish color highlights just how little is known about crayfish, especially the burrowers. Their lives are complex, and biologists are just scratching the surface with our current understanding of their unique lifestyle.

## CHAPTER 3

# Ecosystem Engineers and Keystone Species

## Beavers without a Backbone

BEAVERS GET ALL THE CREDIT. They are heralded for their abilities as ecosystem engineers; by eagerly building their dams, they create a home. And with this feat of animal engineering, not only do they create a cozy home for themselves, but they transform a speedy flow of water into a slow-moving pool—indirectly making accommodations for insects, fish, amphibians, birds, and other mammals. One beaver-built lodge is engineered and the entire local ecosystem benefits. Where beavers occur, animals and plants flourish. Where beavers have been decimated, animals and plants suffer.

By no means am I trying to start a war with the members of the beaver community, whom author Ben Goldfarb refers to as the Beaver Believers. Beavers are amazing, don't get me wrong. But I do want to highlight the often underappreciated crustaceans that, just like beavers, are masters at engineering the environment. Even though a beavers-to-crayfish comparison of ecological importance can't be made, one thing is clear: Crayfish provide invaluable contributions to their communities. Just like beavers and their dams, wherever you have crayfish, you have much more than that. Crayfish are among the most industrious of animals, as they create homes for more or less any creepy-crawly you can think of: frogs, dragonflies, salamanders, and everything in between—both in water and on land.

IF CRAYFISH EVER HAD TO partake in first-day-of-class icebreakers and confess their hobbies, they would say burrowing—whether it's under rocks in a babbling brook or in a poison-ivy-filled ditch. Digging their claws into the ground and moving substrate around is their default setting. Plop a crayfish in an aquarium and you will notice it reorganizing the rocks and gravel to its liking. Come back the next day, and you will find a new arrangement, as if the crayfish is never satisfied. Crayfish have been burrowing for as long as crayfish have existed, well over 300 million years—before, during, and after dinosaurs roamed the earth. At one point, some of the most famous dinosaurs—sauropods, stegosaurs, and velociraptors—certainly munched on a crayfish if they caught one out of its burrow. Some of the best fossil evidence of crayfish doesn't even come from preserved bodies but from preserved burrows. Burrowing is part of what makes a crayfish a crayfish.

For the primary burrowing species, their complex networks of tunnels are where they spend their entire lives—it's where they eat, it's where they mate, and it's where they sleep. It is their home. But sometimes, a crayfish will move on: It may outgrow the burrow, it can search for new resources when those around its burrow get depleted, or it goes out in search of a mate. In these situations, the burrow and all its amenities are left behind—a home without an owner. Free real estate, although never for long.

FROM THE STREAM BANK, a trained eye can tell if crayfish are present without even flipping a rock, because streams with crayfish are littered with burrows. And within these burrows, an abundance of other animals take shelter. With their hefty, robust claws, crayfish do all the heavy lifting, and freeloaders take full advantage.

Fish are just one group that benefits from crayfish burrows. For example, darters are a group of generally finger-sized benthic-dwelling fish that get their moniker from their ability to quickly dart around a streambed. Throughout Appalachia, darters and crayfish are benthic neighbors. Darters

# FRESHWATER REEFS

Throughout a region known as the Highland Rim in Tennessee, Kentucky, and Alabama, the Slender Crayfish (*Faxonius compressus*) is among the most impressive burrowers out there. Slender Crayfish love chert, which are small crystalline rocks that sound like glass when clacked together; few large boulders exist in chert-dominated streams. Lacking large, stable substrate to burrow beneath, Slender Crayfish do something amazing: They create an underground tunnel system within the chert, almost as if they were burrowing on land working with mud.

Prime habitat throughout the Highland Rim will have thousands of rock burrows within a few square meters, all of which are occupied by not just Slender Crayfish but other crayfish species and dozens of fish species, which weave in and out of these burrows as if they were in a game of whack-a-mole. The result is an environment with a jumbled labyrinth of burrows that house many species, akin to coral reefs, with bountiful nooks and crannies for all life to thrive. Having snorkeled in both saltwater reefs and prime Slender Crayfish habitats, I can attest to their similarities. Based on data I have collected in these streams, I estimate the average number of burrows that this species creates in a 5-by-100-meter pool (about the size of two tennis courts) to be over 25,000!

A Slender Crayfish (*Faxonius compressus*). Photo by Zackary A. Graham.

can nuzzle substrate out of the way, but crayfish have a set of massive claws and eight other bulldozing appendages that make excavation easy. Because darters and crayfish cohabit streambeds, these fish get to freeload off crayfish burrows and escape the predator-filled waters above.

The confines of a crayfish burrow can also harbor a collection of small insects and other crustaceans. Aquatic macroinvertebrates (a catchall phrase given to any nonmicroscopic arthropod found in a stream) can be densely packed inside a crayfish burrow: larval dragonflies, damselflies, stone flies, mayflies, caddis flies, black flies, water-penny beetles, and dozens more are only a rock flip away. Starting out as aquatic larvae, many of these animals mature into terrestrial flying adults and serve as a food source for all the more familiar friends: birds, reptiles, and small mammals. Without the safety of a crayfish burrow, many would never survive long enough to make it out of the water and into the mouths of larger animals.

Having so many animals under a single rock, all sheltering from predators, makes it easy for biologists to gauge the health of a stream. Because when crayfish burrows are present, there is usually a healthy community of freeloaders using them. When crayfish aren't present, the lack of burrows and shelters means that many of these animals are out in the open and lack easy access to shelter, leading to an impoverished ecosystem.

Underwater crayfish burrows are an amazing resource. But crayfish burrows on land are a prized possession. For many creatures, finding an empty terrestrial burrow at the right time of year can be the difference between life and death.

A CRAYFISH BURROW in a stream is like a short-term campsite—it gets the job done. But a crayfish burrow on land is like a full-blown mansion, with all of the amenities that many animals need: water at the bottom and a dark place to escape from predators, all housed in a perfect temperature-controlled environment.

Since burrowing crayfish (like all crayfish) require water to breathe, their burrows are an underground oasis beneath the dusty, sunbaked environment above. Some habitats, like the scrubby longleaf pine forests of South Carolina, may not have water in sight for miles, yet burrowing crayfish in the genus *Distocambarus* will dig down several feet to reach the hidden water below. Even if the water table drops, their chamber at the bottom of the burrow can retain moisture and humidity and allow for respiration. In colder months, deep burrows provide shelter from the frost above: Burrowing crayfish will hunker down, waiting to emerge until the warm spring rains. Winter temperatures recorded in burrows show that they stay anywhere from ten

Two types of carnivorous plants, sundews (in the genus *Drosera*; *pictured above*) and bladderworts (genus *Uticularia*), are known to thrive in the presence of crayfish burrows. Because carnivorous plants prefer recently disturbed soil, crayfish burrows are an ideal location for them to grow. Biologists who search for these plants have the best luck in areas with dense congregations of crayfish burrows. Photo by Logan Crees.

to fifteen degrees Fahrenheit warmer than the air directly outside of the burrow, allowing for an ideal location to hunker down in pseudohibernation.

When the soil is in prime condition (after rain, for example), crayfish spend hours or days constructing their burrow, making sure that their chamber is of sufficient size, creating offshoot tunnels, and smoothing down walls without roots blocking the way for efficient movement. With such an amenity constructed in the middle of an often barren habitat, it is no surprise that other animals piggyback off the crayfish's engineering ability. In fact, the most exhilarating part of digging into a crayfish burrow is sometimes not finding the crayfish but having the chance to find other creatures inside.

THE FEAR FOR ANY first-time burrow plunger is often related not to the claws of a crayfish coming out and pinching your finger but rather the fangs of a snake piercing your skin. Although the chances are low, several snake species are crayfish burrow specialists. They spend a greater portion of the year hunkered down in these confined crayfish-constructed chambers.

(Although snakes are known to inhabit burrows, over the tens of thousands of burrows that colleagues and I have dug, I am unaware of anyone getting a snakebite.)

Because burrowing crayfish dig deep enough to avoid the chill of winter above, their burrows make perfect hibernation sites for cold-blooded creatures. One of the rarest snakes in the Eastern United States, the Eastern Massasauga Rattlesnake (Sistrurus catenatus), hibernates in crayfish and mammal burrows. Because snakes, like crayfish, are ectothermic (i.e., cold-blooded, although this description is a misnomer because their blood is not cold), they are capable of being active during only a small window of temperatures. Therefore, a crayfish burrow provides a refuge for a rattlesnake to lie low and wait for spring, when warming temperatures will bring back the snake's activity. With fewer Massasaugas being found each year, management officials recognize the importance that burrowing crayfish play in the rattlesnake's future success. Protecting the crayfish also means protecting much more.

Another fearful and fanged crayfish-burrow-dwelling specialist is the elusive Kirtland's Snake (Clonophis kirtlandii). With a diet of earthworms and slugs, these snakes are adapted to a fossorial, dirt-loving life, but instead of building their own burrows, they rely on crayfish to do the work. Kirtland's Snakes use crayfish burrows not only as hibernation sites but also as a daily retreat.

FOR SOME SPECIES, crayfish burrows are not just a bunker to hide from predators. Some species use burrows to raise their young. One of North America's most endangered insects, the Hine's Emerald Dragonfly (Somatochlora hineana), has been having its preferred wetland habitat replaced by roads and shopping malls for decades. This species lays eggs almost exclusively in crayfish burrows. In Illinois, Michigan, Wisconsin, Missouri, and just across the border in Ontario, Canada, where the Hine's Emerald lives, seasonal refuge from drought and drying has become a serious issue, because there is now a lack of year-round available wetland habitat. During periods of drought, Hine's Emerald larvae leave the drying streams and rely on crayfish burrows as their primary aquatic habitat. Without crayfish burrows for the dragonflies to access, the dry time of year would desiccate hundreds of endangered larval dragonflies. Other insects, such as the solitary wasps known as Cicada Killers (Sphecius speciosus)—use your imagination as to how they got their name—also use crayfish burrows for reproduction. There are probably hundreds of small invertebrate species that use crayfish burrows for some aspect of their life. Crickets, beetles, spiders, worms, slugs, and others are commonly discovered within the confines of a crayfish burrow.

The Hine's Emerald Dragonfly (*Somatochlora hineana*; *top*) and the Mud Salamander (*Pseudotriton montanus*; *bottom*) are just two of many known species that use crayfish burrows for reproduction. *Top*: Photo by John Abbott/Abbott Nature Photography. *Bottom*: Photo by Jake Scott.

# PLANET CRAYFISH

Life is all about perspective. From our perspective, our world is planet Earth. For other organisms, their "Earth" is restricted. For a unique set of animals, their entire world is not Death Valley, Yellowstone, or the Chesapeake Bay but rather the claws, gills, or abdomens of a crayfish. That's right—not only do crayfish create habitats for other animals, but they also harbor dozens of others that spend their entire life in association with the inside and outside of their bodies. To these organisms, their crayfish host is their planet.

Crayfish worms (family Branchiobdellidae) are leechlike worms often found squirming around on a crayfish's exoskeleton. Despite being related to leeches, evidence seesaws between these worms being beneficial, detrimental, or neutral to the crayfish. On a single crayfish there may be a half dozen unique species of crayfish worms.

Inside of the crayfish, a set of crustaceans known as seed shrimps (family Entocytheridae) inhabit the gill chambers of crayfish. These near-microscopic crustaceans can reach densities of over 1,000 in a single crayfish. While describing crayfishes across the Southeastern United States, Horton H. Hobbs Jr. also specialized in these crayfish-loving seed shrimps, because no extra fieldwork was required.

A crayfish worm on the carapace of a crayfish. Photo by Julien Renoult.

A seed shrimp. Wikimedia Commons.

Some of our slimy, wet-skinned friends also use the moist shelter of crayfish burrows to lay their eggs. At least half a dozen salamander species lay their eggs at the base or inside of crayfish burrows, including the Mud Salamander (*Pseudotriton montanus*) and the federally endangered Frosted Flatwoods Salamander (*Ambystoma cingulatum*). Because amphibian eggs require constant moisture, a crayfish burrow is the perfect reproductive refuge.

Management biologists that protect and monitor wildlife populations are aware of the relationships between burrowing crayfish and other animals. Preserving and caring for habitats that promote healthy populations of crayfish can lead to other animals being equally as successful. Crayfish really are the beavers without a backbone.

# Friendly Fire

Fires can destroy an ecosystem. Despite this, grasslands and prairies are purposefully burned to give the environment a "reset" and to ensure that larger woody plants don't take over the tall grassy habitats that many animals need. Open grassy landscapes were once managed by naturally occurring wildfires. Now, humans set them ablaze in controlled burns. During an intentional burn event, most wildlife can escape the heat. Large mammals and birds have obvious means to get away. For smaller, less mobile animals, like amphibians, fires can mean disaster. Such fired fields house one of the most interesting relationships in nature—between a crayfish and an unlikely resident that relies on a crayfish-made burrow for the majority of its life.

In the expansive field of crayfish burrows scattered throughout the Hillenbrand Fish and Wildlife Area in south central Indiana lives a species of frog that has adapted to live eleven months out of the year in and around these burrows: the Crawfish Frog (*Lithobates areolatus*). Crawfish Frogs are medium sized and covered in black or golden circles. Although this species can be found in a widespread area across nearly a quarter of the United States, they are only able to exist where burrowing crayfish live. Without the burrows of the crayfish, these frogs are left homeless and defenseless—free pickings for the snakes, birds, and coyotes that prowl the areas in which they live.

Crawfish Frogs love hiding out in crayfish burrows so much that they have been deemed "the most secretive amphibian in North America" by media sources, because just like burrowing crayfish, they spend nearly their entire lives underground within the confines of a burrow. Over the few past

A Crawfish Frog (*Lithobates areolatus*). Photo by Kory G. Roberts.

A Crawfish Frog (*Lithobates areolatus*) demonstrating a perfect glove-like fit into a crayfish burrow. Photo by Nate Engbrecht.

decades, populations of Crawfish Frogs have been disappearing, with nearly half of their known breeding habitats being impacted by human activity. At the state level, Crawfish Frogs are endangered in Indiana and Iowa. Areas like Hillenbrand are paramount to this species' recovery, as this plot of land holds the highest-density breeding population of Crawfish Frogs in all of Indiana.

A FIRE BURNED THROUGH Hillenbrand for the last time in 2011. The dense prairie foliage became wide-open black, charred fields. And Michael Lannoo, a professor of anatomy at the Indiana University School of Medicine, knew he had a rare opportunity. Lannoo had been studying Crawfish Frogs for a while before the 2011 burn, but he had never been able to get an accurate representation of just how many crayfish burrows, and therefore potential homes for Crawfish Frogs, there were at Hillenbrand. With the recent burn, counting every single burrow on the eighty-acre plot would be possible. This survey would act as a census, estimating the population of both the frogs and the crayfish. Despite his circle-rimmed wire glasses and his medical school association, Lannoo is a hardened field biologist and longtime frog lover. Part of this experience taught Lannoo when to ask for help, as this burrow-counting task could not be done alone.

Together with a pack of eager graduate students, Lannoo set off across the field, stopping every few feet to count crayfish burrows. And because this area harbors the highest-density population of Crawfish Frogs in Indiana, Lannoo knew, without a shadow of a doubt, that there were going to be burrows aplenty. In total, there were 5,603. Combining this data with Lannoo's estimate that around 60% of the burrows in this field would be vacant, in this single plot of land, there could be over 3,000 burrows occupied by crayfish, with the thousands of others serving as empty homes for Crawfish Frogs. A field full of rental burrows, waiting to be reserved.

CRAWFISH FROGS migrate in masses from these burrows to breeding ponds in early spring—moving only during times of ample rainfall. Males travel first to be sure they don't miss the multi-thousand-egg masses that the females lay. Once the eggs get dropped off by the females and fertilized by the males, each Crawfish Frog takes anywhere from a day to a week to get back to a burrow. On their journey, they make pit stops in temporary "rental burrows," or what Lannoo calls secondary burrows. But eventually they make it back to the burrow they call home, the same burrow that they just spent the last eleven months in. Each year, the frogs make this journey. They leave their home, take a few thousand hops to the breeding pond, lay

# CRAWZILLA

At Hillenbrand, there are at least two species of burrowing crayfish. One of them, discussed earlier, is the Digger Crayfish (*Creaserinus fodiens*). But the second species is likely responsible for the majority of burrows in this plot of land: the Crawzilla Crawdad (*Lacunicambarus chimera*)—a crayfish with one of the most epic common names ever. The Crawzilla is unofficially the largest of North America's burrowing crayfish, and it excavates wide but relatively shallow burrows, ranging from 1 to 1.5 meters deep (3.28–5.25 ft). I have been elbow deep in burrows of this species without any part of my arm touching the tunnel walls.

A Crawzilla Crawdad (*Lacunicambarus chimera*). Photo by Zackary A. Graham.

their eggs, and then take a few thousand more hops back to the exact same burrow.

Figuring out that each Crawfish Frog has a single burrow that it returns to each year was no easy task. The investigation started when one of Lannoo's PhD students, Jen Heemeyer, started slapping radio transmitters either on (in the form of a backpack) or into (with a surgical implant) the frogs. Heemeyer spent hours chasing Crawfish Frogs, though in truth, there wasn't much chasing to speak of. Instead, the task was more about meandering across the field with a three-foot-long antenna that looked like a relic from a bygone era of television, complete with forked prongs jutting out at odd angles. One afternoon she tracked a frog, "Number 62," all the way back to the exact same burrow in which it had spent the past year. On this trek from the breeding pond back to the burrow, Heemeyer witnessed Number 62 pass up *hundreds* of crayfish burrows. And instead of settling in any of those, Number 62 made a beeline—or more realistically, a haphazard zigzag frog-line—straight to the exact burrow it had camped out in for the last eleven months.

In Heemeyer's years of tracking the frogs, she found that Number 62's behavior was not out of the ordinary; it was the norm. Each year, each Crawfish Frog returns to the exact same burrow. The longest consecutive use of a home burrow by a tracked frog was *five years*—a long time! With a maximum lifespan of ten years, Crawfish Frogs may spend the majority of their life in a single burrow.

But why would a Crawfish Frog do this? Why would they risk their lives and hop around a grassland back to a single burrow—passing up hundreds of presumably perfectly good burrows on their journey? Lannoo and Heemeyer believe that the answer does not relate directly to the Crawfish Frogs, or to the crayfish themselves, but to the predators that roam the fields in hopes of catching a meal.

If the weather is right and insects are abundant, Crawfish Frogs leave the protection of their burrow. But they don't go far; they stay within a single hop of safety. If they spot a predator, a quick hop into their burrow gives them protection. Then they wiggle themselves around and puff themselves up. When inflated, their rounded snout creates a perfect seal to the burrow, like a manhole cover. The snug fit of the Crawfish Frog in the burrow makes it impossible for any predator, whether snake, raccoon, or coyote, to get them out. Lannoo believes that a Crawfish Frog needs a perfect-fitting burrow to be protected from predators. Once a Crawfish Frog finds one that fits, the burrow becomes a lifelong refuge and is worth passing up hundreds of potentially loose-fitting burrows on its way back home.

# PARTY DOWN UNDER

Every so often, Lannoo would come across what he calls a "party burrow"—a subterranean hangout where multiple tunnels converge into a single chamber and, improbably, different species coexist like guests at an awkward mixer. The most frequent partygoers Lannoo saw were Crawfish Frogs (*Lithobates areolatus*) and juvenile crayfish. Occasionally he found a Crawfish Frog sharing a burrow with a garter snake. But the pièce de résistance came from an observation where all three species—frog, crayfish, and snake—turned up together in the same burrow.

"Crawfish burrows to Crawfish Frogs are what shells are to turtles," says Lannoo—without them, they are defenseless. On a year-to-year basis, an estimated 65% of young Crawfish Frogs make it to the next generation thanks to crayfish burrows. Other frog species' survival rates are closer to half that, making these burrow specialists true survivors. But sadly, the number of Crawfish Frogs across the country is declining, and the similarly declining populations of crayfish surely play a role. Across their wide range, Crawfish Frogs likely use the burrows of well over a dozen different crayfish species. A trend becomes clear: When you provide care and protection for burrow-building crayfish, you protect not only their homes but also the entire ecosystem that relies on these burrows.

## A Dinner Fit for a Queen

Winter brings simplicity for stream crayfish. They spend the cold months in a burrow beneath a rock and below the ice. When cold weather approaches, crayfish hunker down for the season in a state of inactivity. They are unlikely to move or eat during this time. But as temperatures rise, the warmth means that crayfish will feast on anything and everything: leaf litter, aquatic plants, and whatever animals can be found—it's all on the menu.

In a full-on feasting frenzy, crayfish pack in nutrients after their bout of inactivity. If a crayfish doesn't get enough food during this time, it falls behind the competition, who move on to bigger and better exoskeletons. A crayfish that can bulk up during the advent of warm weather will be bursting at the seams. It's been the exact same size for more than half a year—but not for long. It is time to grow. Time to molt. In the long term, molting allows crayfish to grow and eventually become better at fending off predators and acquiring mates. In the short term, each molt puts the crayfish in a state of extreme fragility, making it susceptible to predators looking for a soft-shell dinner.

Unlike human bodies (which have endoskeletons), crayfish skeletons are on the outside of their bodies (i.e., they have exoskeletons) and the soft tissues and organs are under it. Most of the year, the calcium-rich exoskeleton is hard and serves its protective duties. But seasonal changes in water temperature, moon cycles, and food availability all trigger a cascade of hormonal events that occur beneath the exoskeleton. The result is the formation of a new, larger exoskeleton that grows underneath their old one.

Just like you would if you outgrew the waist on a pair of pants after a winter bulk, crayfish seek the refuge of a burrow. It takes them anywhere from a few minutes to an hour to squirm and wriggle around to bust out of their old exoskeleton and emerge in their new one, leaving behind a shell of their old self.

When molting, the junction at the carapace and abdomen bursts open and then the crayfish must squirm its way out, its new set of claws being the last to emerge. In the process of growing a new exoskeleton, past injuries heal and lost appendages slowly grow back to their original size after several molts.

In temperate regions, most crayfishes will molt once in spring and again in late summer (although there is plenty of variation here). Molting season is an exciting time to be in the field as a crayfish biologist. When molting, crayfish remove up to a year's worth of gunk and grime from their bodies. They replace their old exoskeleton with a clean, vibrant coat of armor. Just like how car enthusiasts appreciate a sparkling car wash, crayfish are well appreciated when they are at their cleanest. A recently molted crayfish has yet to reacquire the calcium that hardens its exoskeleton, so it is strangely gelatinous to the touch and needs to be handled with care.

To put this helpless, gooey time in their lives behind them, the crayfish must find calcium to harden back up and be protected once again. Calcium can come from food or be absorbed from the environment, but crayfish also have two secret stashes of calcium, called gastroliths, that serve as a bank of this rock-hard element. All crayfish have gastroliths in their cardiac stomach, which they use to store calcium before they molt. Immediately after a molt, they absorb calcium from the gastroliths to give them a head start on the path to hardening. A recently molted crayfish will also consume its old, ghastly exoskeleton for another quick boost of calcium. But crayfish

Crayfish molting sequence. Photos by Chris Lukhaup.

will never consume their claw tips, which are the hardest, and presumably most difficult, part to chew.

Some crayfishes, like ones in the genus *Faxonius*, go through mass molting events; you can arrive at a stream and find hundreds of ghastly crayfish exoskeletons looking like a massive die-off just occurred. But flip a rock, and you will find an abundance of soft, recently molted specimens waiting for their exoskeletons to harden. Until that happens, soft-shell crayfish are a delicacy in the natural world. They are an easy meal that puts up little fight, like a shelled pistachio. Sometimes, these crayfish wave their claws around and behave like they can still put up a fight, but during this time they are unable to generate a significant pinch despite their bravado. And when these helpless crayfish are available, it becomes open season for predators that rely on these gelatinous masses of protein.

WITH THE SCENT of freshly molted crayfish in her Jacobson's organ, a small, nondescript, nonvenomous snake known as a Queen Snake (*Regina septemvittata*) nudges the rocks in the streambed on the hunt for a soft-shell meal. Her highly sensitive scent organ, named for Danish anatomist Ludwig Levin Jacobson, is homed in on the crustacean's scent. Paired with the speedy in-and-out flicking of her tongue, she knows her next meal is close.

A Queen Snake (*Regina septemvittata*). Photo by Kory G. Roberts.

No larger than a typical garter snake and having poor eyesight, Queen Snakes are masters of picking up on underwater scents. Their reliance on such scents is taken to an extreme, as Queen Snakes exclusively consume recently molted crayfish, for breakfast, lunch, and dinner; they are astacivores, animals whose diet consists entirely of crayfish.

Immediately after a crayfish molts, the water becomes polluted with a molting hormone known as ecdysone. To a Queen Snake, ecdysone is like catnip, sending the snake on a feeding frenzy. Countless experiments have been conducted with these crayfish-crazed reptiles. You can put the scent of a recently molted crayfish on just about anything, and Queen Snakes will lunge, attack, and wrap around it, hoping it's the real deal. Because Queen Snakes are semiaquatic, they slither under rocks in a stream in search of helpless, recently molted crayfish. Without a set of masticating teeth or a strong body for constriction, Queen Snakes consume their meals whole, always swallowing tail first and claws last—making sure the crayfish's business end cannot cause harm. If the soft crayfish puts up a fight, the Queen Snake picks off its claws first, setting them aside and rendering the crayfish even more defenseless.

Ranging from the Great Lakes region to the Florida panhandle, wherever these snakes live, you are guaranteed to find the crayfish that they rely on. Besides the Queen Snake, an astacivorous diet has evolved two other times in snakes! These "crayfish snakes," though distantly related, all converged on a similar diet based on the sheer dominance of crayfishes in their habitats.

The Striped Crayfish Snake (Liodytes alleni) inhabits swamplands throughout Florida and south Georgia. With their larger body size, Striped Crayfish Snakes are robust enough to take on both recently molted and normal hardened crayfish prey, which they wrap their bodies around in order to control the crayfish. Similarly, the Glossy Crayfish Snake (Regina rigida) is found throughout the coastal plain in the Southeastern United States and also consumes a mixture of hard- and soft-shelled crayfishes. The last crayfish snake, the Graham's Crayfish Snake (Regina grahamii)—sadly this species was not named after one of my ancestors—also consumes soft-shell crayfish almost exclusively, and it relies on this ephemeral resource for its entire life.

Because Queen Snakes and Graham's Crayfish Snakes are specialists in consuming recently molted crayfishes, it's open season for them whenever molting season occurs. During these times, the snakes are highly active and need to consume as many crayfish as possible. Within a few short weeks, their preferred food source will harden up—capable of defending itself with its claws, making it a losing battle for the snakes.

Wherever these snakes are present, so are the crayfish they need to consume. When crayfish are absent, the snakes may disappear. And if the snakes disappear, heron and raccoon populations lose one of their food sources. The connectedness of an environment is ever present when you consider the role that just one animal, like a crayfish, plays, as it is often right in the center of it all.

## Stuck in the Food Web

"Crayfish eat everything, and everything eats crayfish," says every crayfish biologist ever. This adage is stated wherever crayfish biologists go—whether they are talking to a third grader at a science fair, a wildlife officer, or an educated bunch at an international conference—because the importance of crayfish goes far beyond their role as ecosystem engineers. They also serve as keystones in the center of food webs, both eating and being eaten by both land and aquatic animals. They are positioned at the center of it all. Keystone species like crayfish have a disproportionally large role in ecosystems.

Unlike the highly specialized diet of crayfish snakes, which exclusively eat off the seasonal specials menu, crayfish aren't picky. Crayfish are generalist, opportunistic omnivores. They will eat anything at any time. At one point, crayfish were lumped into a category known as shredders. Shredders such as caddisflies and mayflies eat by shredding the plant material that falls into a stream, which helps break down the nutrients into more readily available energy forms. The historic perspective that crayfish are shredders perpetuated the wrong idea in the community that "a crayfish is a crayfish" (meaning that every crayfish behaves and acts the same in an ecosystem)—disregarding the fact that there are more than 700 crayfish species and they have variable diets and highly diverse environmental roles.

To extinguish the idea that "a crayfish is a crayfish," thousands of crayfish stomachs have been dissected to understand what these crustaceans eat. This is a tough job, as crayfish have a gastric mill—a pair of hardened teeth that grind their food—leaving nothing but bits and pieces. Still, plants, roots, whole earthworms, slivers of beetle exoskeletons, and fish scales are just some of the stomach contents that can be identified under a microscope. Gut dissections are now paired with a chemical analysis known as stable isotope analysis, which determines where crayfish are getting their nutrients and their position in the food web. In combination, these methods have revealed how variable crayfish diets are. Their diets change throughout their lives, they change based on the season, and they can even vary from species to species.

As freshly hatched craylings, these little-clawed rascals are ravenous meat eaters, with studies showing that nearly 90% of their diet comes from animal protein. Craylings consume smaller crustaceans (occasionally their siblings) or larval insects in huge quantities. Because protein is required to grow, juvenile crayfish try to consume as much of it as they can as early in life as possible so they can reach a size at which they are less susceptible to predators.

As crayfish molt into adulthood, their diet diversifies. Seasonally, if there is a time of year when macroinvertebrates increase in number, the crayfish turn to a more predatory lifestyle, actively foraging at the base of their burrow. If insect prey dies down, and leaves start to drop, they switch to a shredder diet. The leaves themselves are not exceedingly nutritious on their own, but the microscopic phytoplankton and zooplankton that inhabit them are, making leaf salad a high-calorie meal. If plant and animal resources dwindle, crayfish turn to feeding on inedible small rocks and sand, which, like the leaves, contain microscopic nutrients to graze on.

I learned about the flexibility of crayfish diets early in my career. To have crayfish available for my PhD research, every two weeks I'd take a trip to trap crayfish in man-made lakes throughout Arizona. I was a tall, gangly figure hurling traps into highly populated lakes, so bystanders loved to snoop around and ask what I was doing. After telling them I was looking for crawdads (see the map in chapter 1), I became a human suggestion box. Bystanders loved to rattle off what type of bait their great-great-grandpappy had used to catch crawdads in his traps: dog food and cat food were most suggested (dried or canned), followed by bread, raw bacon, fish, hot dogs, chicken livers, chicken gizzards, chicken necks, leaves and twigs (for the ecologically minded), and if you are feeling wild, marshmallows and bananas (separately or together). Some of these suggestions went on to be tested in after-school science clubs I led for local Arizonans, as students tossed just about anything into a crayfish aquarium, and unsurprisingly, the answer to the age-old question of "What do crayfish eat?" is an easy one. They eat just about anything.

But biology always has exceptions to the rule—outliers exist. Researchers have shown that the Bottlebrush Crayfish (*Barbicambarus cornutus*) is a top predator in the streams it inhabits. Both gut content and stable isotope data demonstrate that this species has a role in the food web closer to that of a predatory fish than that of other generalist crayfish. Its heavy reliance on predation and animal protein may relate to its ability to reach such large sizes. In contrast to this protein gigantism hypothesis, however, the largest species of crayfish in the world, the Tasmanian Giant Freshwater Crayfish (*Astacopsis gouldi*), has taken a different route to reach massive sizes, consuming almost nothing but decaying logs and plant material.

The historic idea that, dietarily, "a crayfish is a crayfish" has been overturned by numerous demonstrations of the diverse foods that crayfish consume. Depending on the stream, species, age, and season, crayfish consume and process all types of nutrients. A single crayfish may be tied to hundreds of different food sources throughout its life. But remember that when dealing with food webs, the organisms found in the middle of a web aren't just predators; they are also prey.

BOB DISTEFANO, a longtime astacologist and member of the Missouri Department of Conservation, faced a lot of departmental hesitancy for supporting the ecological importance of crayfish when he first got his start as a crayfish conservationist. DiStefano has years of firsthand experience getting people to care about (and pay attention to, both mentally and financially) crayfish. Crayfish are often outshined by fish and the anglers that bring in money buying licenses year in and year out. Both state- and federal-level conservation agencies have a long history of pouring money into research and work related to preserving habitats for fish and learning about their biology. But only recently, much thanks to DiStefano, people started paying attention to crayfish.

To demonstrate the importance of crayfish to his colleagues, DiStefano compiled a list of every known animal that consumes crayfish. He figured that if he could show just how many animals rely on crayfish for food, it would clearly demonstrate their ecological importance. In his research, he found documentation of 208 different animals that consume crayfish: five insects and arachnids, seventeen amphibians, twenty-nine mammals, fifty reptiles, sixty birds, and sixty-five different types of fish. And that was just what was reported in the literature and observed by biologists. Due to crayfishes' often high abundance, there are likely hundreds of other organisms that rely on crayfish for food.

Fish came out on top in DiStefano's list of crayfish consumers, with many common species being known to eat crayfish: catfish, gars, chubs, carps, darters, and bluegills. You name it, crayfish get eaten by them. Out of all the species, bass are probably the most reliant on crayfish, with nearly every species of bass being a clear crayfish lover: Studies looking at stomach contents of bass have found that anywhere from 60% to 80% of bass eat nothing but crayfish.

Some of the many animals known to consume crayfish. Photos by Robert So
(*top left*), Bernie Duhamel (*top right*), and Anne Fraser (*bottom*).

# NUTRIENT CYCLING

Protein has to come from somewhere. It doesn't magically appear in the bodies of an ecosystem's top predators, like birds of prey or mountain lions. It all starts with smaller, less recognized sources of protein, like crayfish. By consuming large amounts of unused organic material in an ecosystem (like leaf litter), crayfish are able to harness nutrients to construct their own muscles for growth and movement. And because crayfish are often the most abundant large invertebrates in a stream (or terrestrial) ecosystem, they serve as a source of protein to larger predators. This places crayfish in a key role as nutrient cyclers. Crayfish convert unused organic matter into usable protein for larger predators. Crayfish consume leaf litter. Fish eat the crayfish. Bears eat the fish. Bear poop decomposes in the soil. Plants harness these soil nutrients. Plants produce leaves, which fall into streams. And repeat.

DiStefano credits his lengthy list of crayfish consumers as the reason government agencies started thinking about crayfish populations in the same way they do fish. With such a massive list, the ecological impact of a single crayfish population can be immense. It's not just humans that enjoy a crayfish feast (see chapter 6), but so do hundreds of other animals, both on land and in water.

CRAYFISH GET EATEN by hundreds of animals, eat nearly everything, serve as hosts for dozens of hitchhikers, and serve as an indicator species of the environment's health, all while creating homes for ecosystems both on land and in water. Without them, North American ecosystems would be unrecognizable. Soon after crayfish populations diminish (see chapters 7 and 8), many other animal populations follow suit. As a keystone species, crayfish play a critical role in the ecological theater in which they have been cast. They were around when *Tyrannosaurus rex* was at the top of the food web, when Pangea split, and during the recent 60 million years of mammal dominance. Humans are the newcomers and are just now beginning to appreciate their importance.

# CHAPTER 4

# Crayfish Brains and Crayfish Benders

## Laboratory Wonders

WITH A CLOSETFUL of metal wire racks stacked to the ceiling and a few fold-up tables against the walls, this dusty, narrow closet is ready for research. Spread evenly across the multitiered wire racks are 120 plastic take-out containers—each filled with an inch or two of dechlorinated water. Crayfish are easy to please and do not mind cobwebs or rattling pipes overhead (at least they never complain about it). Although this setup seems worse than your average roadside motel with a number in its name, as long as the crayfish have water and a dark, secluded space, they are content. In the wild, crayfish spend most of their time in similar conditions, tucked under a rock. This cramped setup mimics their reclusive lifestyle.

Early in my PhD studies, Arizona State University's biology department gifted this glorified closet to me and my adviser, Dr. Michael Angilletta, so that I could conduct my dissertation research in a "controlled laboratory environment." Angilletta's typical research tries to understand how flies, lizards, and other animals adapt to changing temperatures. His normal lab space is well equipped for this type of work, with hundreds of thousands of dollars' worth of incubators, calorimeters, respirometers, and other fancy devices. But after I convinced Angilletta to let me study crayfish behavior (it wasn't a hard sell—he sensed my conviction), we realized I would need a new space for my work.

"All I need is somewhere to house a hundred crayfish or so, some tables, and a sink. Even a closet would work," I told him. And in only a few months, my wish was granted, although in a more literal sense than I thought. I was gifted a decrepit storage closet that became my research home for the next several years. This closet marked a turning point in my PhD journey—a journey that, up to that point, had felt more like an endless literary scavenger hunt than actual research. For the first two years, I had been essentially living in a universe of books, spending weeknights and weekends alike buried in literature, amassing more questions than answers. My weekly meetings with Angilletta seemed only to thicken my uncertainty. But after being granted a research space—a closet, but a space nonetheless—I started to picture the experiments I wanted to conduct over the following years.

Most people picture scientists in fancy laboratories, wearing white coats, toying with expensive instruments, with Bunsen burners and Erlenmeyer flasks lining the clean, orderly space. This might be true for, say, chemists and microbiologists. But a lot of research is conducted in makeshift laboratories by biologists wearing sandals (typically a weathered pair of Chacos for the ecologist nature-hugger types or Birkenstocks for a more sophisticated look), who have created their own jerry-rigged scientific devices. It's safe to say that no one wearing a lab coat has ever stepped foot into my closet filled with crayfish in plastic take-out containers, my "controlled laboratory environment."

In the former closet, now laboratory, I planned to study animal communication and aggressive behavior. In the same way that humans engage in yelling and pushing matches before throwing fists in a bar fight, nonhuman animals often communicate with their opponents to minimize the risk of harm. Because fighting is an everyday affair for many animals in nature, I was designing experiments to learn how these animals communicate before a fight.

I want to clarify my research question above. Notice how the study plan does not mention crayfish. At its core, my research question had little to do with crayfish. I was interested in learning about broad, generalizable patterns relating to animal communication and aggression—and not crayfish specifically. I was simply using crayfish as a *model organism* to investigate these questions. In this way, the results of my experiments could be extrapolated to glean information about fighting behavior and communication in all animals—from beetles to chameleons.

Instead of using crayfish, I could look into the same question by using Komodo dragons, elephants, or gorillas—but you can imagine how these animals may not be as easy to work with as crayfish (they certainly wouldn't have fit in my closet laboratory). This is the benefit of working with crayfish, the favorite research animal for of all kinds of biologists: Most species are easy to obtain and simple to keep alive, and importantly, they exhibit natural behaviors in laboratory settings. In Arizona, working with crayfish was as simple as getting a scientific collecting permit and dropping off baited traps in various recreational lakes. Each trap would yield thirty to forty crayfish, which made boosting the numbers for my experiments simple. Getting enough animals for a project on any of the other large-bodied animals listed above leads to obvious issues and is why biologists focus their work on model species.

## Crayfish: America's Next Top Model (Organism)?

Model organisms are cornerstones of biology research. Many of the most important discoveries about physiology, genetics, and evolution all started with model organisms. Depending on what type of questions you are interested in answering, you can peruse an online catalog of models from biological supply websites, where they can be purchased in the thousands. In just a few days, you can have a plethora of model species shipped to your front door, ready for experimentation. Zebra Fish (*Danio rerio*) have transparent embryos that develop naturally in the laboratory, allowing developmental biologists to study how organ systems develop. Without Fruit Flies (*Drosophila melanogaster*), our understanding of genetics would be far behind what it is today. And because humans are mammals, physiologists and medical doctors use House Mice (*Mus musculus*) and Brown Rats (*Rattus norvegicus*) to learn about how our bodies work.

Nestled somewhere among the ranks of top model organisms lies the crayfish. Behavioral biology, ecology, and neuroscience all rely on crayfish, primarily the Red Swamp Crayfish (*Procambarus clarkii*). In these disciplines, the willingness of a crayfish to behave naturally in a laboratory setting is what makes them unique. If you put a crayfish in an aquarium, it will start exploring. If there is substrate in that aquarium, it will start digging or burrowing. If you add another crayfish to that tank, they will fight. If that crayfish is of the opposite sex, they will mate. Other model animals take a while to acclimate to lab conditions. But not crayfish.

Crayfish were thrust into the limelight as a model organism by the famous advocate of Charles Darwin's evolutionary theory, Thomas Henry Huxley, in his 1880 book *The Crayfish: An Introduction to the Study of Zoology*, in which he uses crayfish as an example of how one can study a specific organism and learn about diverse zoological and biological principles. In the preface of this text, Huxley states that his goal was not to prepare a comprehensive monograph on crayfish biology but rather "to show how the careful study of one of the commonest and most insignificant animals, leads us, step by step, from every-day knowledge to the widest generalizations and the most difficult problems in zoology; and indeed, of biological sciences in general." Although crayfish are far from "insignificant" (I full-body cringe every time I hear this word in relation to crayfish), Huxley was the spark for researchers to start thinking about crayfish as model organisms in biological studies.

More than 150 years later, researchers are doing exactly what Huxley advised: using crayfish to tackle some of the biggest problems in all of biology. Researchers are now studying crayfish in ways that perplex outsiders (and sometimes even me), including studies that shed light on decision-making, drug addiction, human behavior, and how our nervous system controls it all.

## Giant Tails with Giant Neurons

The human brain is among the most puzzling biological systems. Our brain accounts for, on average, 2% of our body weight but nearly 10% of our body's metabolism. It takes a lot of energy to fuel the tasks that are possible seemingly only in humans: playing sports, saving for retirement, or writing books about crayfish. But our brain can't do it all alone; it requires a supporting cast. Our brain is connected to the spinal cord, which runs down our back, and connects to trillions of smaller nerves that spread through the rest of our body.

Even with big, complex brains, humans are still surprisingly clueless when it comes to many areas of neurology—like how memories are stored or how regions of our brain communicate. These questions have no simple answers. But answering them is possible; it just requires years of careful long-term studies and an understanding of neurophysiology. Doing research directly on humans is typically off-limits, so scientists must find other models to use in our research. If animals have a backbone (i.e., they are vertebrates), like Zebra Fish, rats, monkeys, or humans, there are strict guidelines on what scientists can and can't do with them in an experiment. But for invertebrates—animals that lack a backbone, like Fruit Flies, earthworms, and crayfish—there are no guidelines or governing bodies that limit work with these animals. In the United States, animals without a backbone do not require specific permits or permission to experiment on. (The intelligent cephalopods, the group that contains octopuses, squids, and cuttlefish, are an exception to this rule.)

Despite their "buglike" appearance, crayfishes share the same underlying neural architecture as other complex animals, including humans. With research on invertebrates being the scientific Wild West, researchers are free to gather as many crayfish as they want and conduct any number of experiments. And experiments with crayfish have led to some of the biggest brain-related discoveries of the past century—something Huxley would be proud of.

THE NEURAL BASIS of most animal instincts has been well studied in crayfish. Since crayfish are seldom at the top of the food chain, they need to be able to fend off and escape from predators. If a crayfish gets caught in the open, far from shelter, it has two options: fight or flight (not literal flight, but they have a similarly effective escape mechanism, detailed below).

When it comes to the crayfish fight response, it's all about their claws. Faced with a predator, a crayfish will brandish its claws in front of itself, hoping to deter the predator by flaring its claws out into a posture known as a meral spread. This behavior is the crayfish equivalent of a human throwing their hands up and saying, "You wanna mess with me? Look how big I am!" In my PhD research closet, when I walked down the rows of research subjects in my makeshift laboratory, each crayfish would throw its claws up in a meral spread—attempting to intimidate me (it rarely worked). Experiments show that crayfish with larger claws are better at deterring predators than crayfish with smaller claws, and predators preferentially prey on smaller-clawed crayfish for this reason. But only certain well-endowed species can successfully fend off predators with their claws. For most crayfish, their first instinct is to flee, and luckily, crayfishes have a Houdini-esque ability that they use to evade capture from any predator, including a crayfish biologist.

When left with no other option, crayfish engage in aquatic flight by rapidly flexing their massive abdomen (what you think of as the tail) muscles and eliciting a tail-flip escape response, which sends the crayfish darting backward and upward away from danger. If you have ever eaten crawfish or

The meral spread threat display. Wikimedia Commons.

The tail-flip escape response. Photo by David D. Yager and Jens Herberholz, University of Maryland, College Park.

their marine cousins, the succulent abdominal muscles are what you drench in butter and lemon before consumption. Nearly half of the crayfish's body is its abdomen, and this is its greatest asset in escaping from predators.

Imagine you're nestled under a rock, and you haven't moved in four months. All of a sudden, a giant hand from above lifts up your home and starts thrusting around in the muddy water trying to get a hold of you. You know your claws are no match for these huge meaty appendages, so you use your only other resource, your tail, to quickly jet away and live another day.

During a tail flip, the crayfish rapidly squeezes its hearty tail muscles, which creates a force so strong that it sends it flying backward to safety. To create even more force, the crayfish spreads out the tail fan at the end of its abdomen, which increases the power generated from the thrusting escape response. A tail flip can be triggered in 30 milliseconds (0.03 seconds; thirty times faster than the blink of an eye) to allow these crustaceans to jet away to safety.

Early neurobiologists took an interest in how crayfish tail flips work. Soon after, biologists began shoving electrodes into crayfish and measuring what was going on underneath the carapace. With precisely positioned electrodes, you can send an electrical signal straight into the crayfish's nervous system, which allows biologists to try to figure out what area controls what behavior.

After only a few years of electrode shoving, the neural underpinnings of the crayfish tail flip became the best-understood neural circuit in the world. Much of this work was pioneered in experiments by Cornelis A. G. Wiersma at the California Institute of Technology in the mid-twentieth century. Born in the Netherlands, Wiersma had an affinity for electric systems that went beyond his crayfish studies. He was a regular at Los Angeles County General Hospital, where he conducted studies of how electronarcosis (an intense form of electrotherapy) could treat schizophrenia.

Before research was conducted on crayfish, the assumption was that the nervous system's influence on the behavior of animals was exceedingly complex. It was thought that crayfish behaviors such as burrowing, fighting, and fleeing would be impossible to untangle from each other, just like that cluster of wires behind your television. But after experimenting with crayfish, Wiersma showed that the tail-flip response was remarkably simple and controlled by a *single* meeting of nerve cells, called neurons. Get that electrode in the sweet spot right at the correct junction of neurons, zap some electricity through there, and the crayfish goes tail-flipping away. It's that simple.

After Wiersma's work showed how a single connection of neurons can control a complex behavior, a storm of other scientists started tinkering

# THE CRAYFISH NERVOUS SYSTEM

The crayfish nervous system is made of the same cellular material as our own. But unlike our system, the nervous system of a crayfish is decentralized, which means that instead of the majority of neural activity occurring in a single area (the brain in vertebrates), the crayfish spreads out its nerve cells. This does not mean that crayfish do not have a brain. They do. It just means that the brain is not the be-all and end-all for the nervous system like it is in humans. The bundle of nerves that forms a crayfish's brain is tucked right between its beady eyes. And running down the length of its stomach is the nerve cord, which looks like a long, stretched-out rope with dozens of large knots tied sporadically down the length of it. The bundled knots are smaller, mini-brain-like clusters of nerve cells called ganglia (singular "ganglion").

Because crayfish ganglia are relatively large and can be seen with the naked eye, neurobiologists have been studying them for decades. In many ways, scientists know more about the inner workings of the crayfish nervous system than they do the human one.

around with the inner workings of the crayfish nervous system. The system was so simple: The crayfish receives sensory input from a predator through vision or smell, the stimulus is integrated through a single neuron, and then the motor output takes action and the abdominal muscles contract, propelling the crayfish to safety with a tail flip. But what is interesting is that crayfish don't just go tail-flipping at every stimulus; they are selective, carefully weighing the costs and benefits. To tail-flip or not to tail-flip? That is the question.

A beautifully designed set of experiments by graduate student Kirstie Bellman and Professor Franklin Krasne at the University of California, Los Angeles, demonstrates the decision-making process in crayfish. Juvenile Red Swamp Crayfish were kept in a lab and assigned to one of three groups. The first group received a small piece of food (a chunk of chicken liver) that was roughly one-third the size of their body, like a toddler lumbering around with a watermelon. The second group was given a huge portion of food, roughly three times the size of the crayfish's body (the entire liver), like a toddler now impossibly carrying an adult-sized beanbag. The third group was not given any food. I call these treatments the small-food treatment, the large-food treatment, and the no-food treatment, respectively.

Each crayfish was randomly assigned to one of the three treatments. Then, during a trial, experiments used a threat stimulus (a run-of-the-mill fishnet—more proof that scientists aren't all about fancy equipment) to slowly approach the crayfish, which simulates an approaching predator. The goal was to trigger a tail flip and see how the crayfish behaved differently based on which treatment they were in. With this carefully designed experiment, the researchers wanted to know if crayfish change their behavior based on the costs and benefits of a scenario—much like a human would.

Here is what they found. In the no-food treatment, the crayfish sensed the net threat and 70% of the time their neurobiology worked exactly as you'd expect, sending them flying away with a few tail flips. The other two trials are where things get interesting and where decisions needed to be weighed.

The large-food trial creates a dilemma for the crayfish. Because the food resource is massive, feeding and escaping are incompatible. Crayfish need to decide whether the risk of losing the food or the risk of a predator is greater. In the large-food trial, when faced with the predator stimulus, a majority of the crayfish chose not to flee and stayed with the food. Only 38% of crayfish in the large-food treatment tail-flipped and escaped the predator. In nature, the remaining 62% would be gobbled up.

In the small-food treatment, the decision was more straightforward. When the crayfish were given a small piece of food, they could both escape the predator *and* feed. The food source was small enough that the crayfish could grab on to the chunk of liver and simultaneously tail-flip away from the net threat. Much to Bellman's and Krasne's surprise, these crayfish escaped with the food an astonishing 93% of the time.

Compare this result to the results of the no-food trial, and you will notice that somehow, someway, having a food source actually *increased* the likelihood of escape (70% chance of escape without food, but 93% chance of escape with food). At face value, these results make little sense, because having to lug a food source around while tail-flipping should burden the juvenile crayfish. But instead, more crayfish successfully escaped when they were carrying a small bit of food than when they had no food.

Bellman and Krasne shed some light on these results in their paper. The crayfish must be able to adaptively lower their neurological "threshold" that triggers a tail-flip response, but only in specific scenarios, like when they are feeding on small (but not large) pieces of food. In the small-food trial, the threshold at which to tail-flip must be easily triggered, because the crayfish brain presumably knows that it has a resource it can escape with. But in the no-food and large-food trials, the neurological threshold does not reach the lowering point and tail-flip responses are less common.

This result makes evolutionary sense, as both escaping from predators and maintaining a food resource is beneficial for survival and reproduction. Overall, these results demonstrate complex neurological decision-making. Crayfish likely evolved to weigh the costs and benefits of every scenario they face—a trait once thought to be reserved only for "higher animals" such as mammals.

From the outside, crayfish are often perceived as bland, mindless drones. But experiments like this demonstrate that these animals are capable of altering their behavior based on context—a surefire sign of behavioral complexity. The results of this experiment were among the first demonstrations of complex adaptive behavior by an invertebrate. Bellman and Krasne were rewarded with publication of their work in the prestigious journal *Science*, under the title "Adaptive Complexity of Interactions between Feeding and Escape in Crayfish." This work communicated to the scientific community that the nervous system of invertebrates was not so simple and that their behaviors could be coordinated, planned, and mapped out. This study proved to be foundational in our understanding of the links between neurobiology and behavior. And it all came about thanks to crayfish.

Studies of behavioral complexity in crayfish were just the start of things. Such work set the stage for countless researchers to realize the benefit of studying crayfish in laboratory settings. Now, experiments with crayfish could be taken to the next level, because neurobiologists had an animal with a well-understood nervous system at their disposal. The obvious next step was to see what happens when crayfish are given drugs.

## Under the Influence . . . Underwater

Have you ever wondered what happens when you give a crayfish drugs? I assume not. But for many biologists, administering drugs to crayfish is as routine as a morning cup of coffee. Caffeine, cocaine, methamphetamines, MDMA (ecstasy), morphine, alcohol, and a slew of antidepressants have all been given to crayfish. By now, hopefully you can see the benefit of these mad-scientist methodologies. Humans know surprisingly little about how many widely consumed drugs actually work, whether they get us high or help us heal. Crayfish provide a stepping-stone to understanding the processes that occur in the human brain when substances that alter the nervous system are swallowed, smoked, or injected.

# CRAYFISH ON DRUGS

When crayfish are given drugs, either by interacting with the diluted drug in water or by direct injection, they start to express stereotypical behaviors. When given cocaine, crayfish will stiffen their bodies, flex their legs, and hold their large claws in a downward manner for a few minutes, as if they must acclimate to the short-lived high. Movement is also heightened during cocaine injections: Crayfish with a cocaine high will pace around their aquarium and constantly sway their antennae as if they are repeatedly investigating a brand-new environment. Following amphetamine injections, crayfish will incessantly pace around the perimeter of their aquarium until the drugs wear off. Being able to identify discrete behaviors associated with drug use allows scientists to understand exactly how the drugs are working.

ONE EVENING, Jens Herberholz, a neuropsychologist at the University of Maryland, decided to pour some research-grade alcohol into a crayfish aquarium, set up a camera, and then come back the next day to see what happened. Herberholz was inspired by the stacks of literature published by Dr. Robert Huber and the work he conducted at Bowling Green State University. Huber was a key proponent in the use of crayfish as a model organism for testing the behavioral and neurological effects of drugs, because crayfish will readily consume, acclimate to, become addicted to, and even face withdrawal from all the same drugs that humans will. Huber focused more on studying "hard drugs" like morphine or cocaine. Herberholz decided to look at alcohol.

The day after he set up his test run, Herberholz came back to witness his crayfish seemingly blacked out from the alcohol: lying on its back and flailing all ten limbs around in the air. That day, Herberholz found out that drunk crayfish act surprisingly similar to drunk humans. Wondering how the crayfish ended up like this, Herberholz went back to watch the footage he had recorded. What he saw mimicked a scene repeated nightly at the local University of Maryland bars. Only fifteen minutes after the alcohol was put into the aquarium, the crayfish started acting strange and a little goofy, stretching its legs and raising its body high off the ground as if it were on stilts. This posture typically occurs during aggression, when crayfish are trying to intimidate each other or when they defend themselves from a predator. But this tipsy crayfish was exhibiting this behavior in a completely isolated tank, without anyone or anything in the laboratory besides the video camera overhead.

Twenty-five minutes after the alcohol was in the container, the tipsy crayfish started to explore its surroundings. During its exploration, the crayfish would spontaneously tail-flip several times in what seemed to be an uncontrollable manner. Herberholz rationalized this by assuming that the crayfish became so intoxicated that every little movement, reflection, or change in the water current it was sensing triggered a tail flip. Ten minutes after the spontaneous alcohol-induced tail flips started to occur, the crayfish ended up on its back, flailing, just like Herberholz found it after prolonged exposure to alcohol.

Just like Huber had in his research, Herberholz witnessed a set of easily observable behaviors when the crayfish was given alcohol. With this information, Herberholz and his graduate students Matthew Swierzbinski and Andrew Lazarchik started to take things to the next level. Humans actually don't know much about how alcohol influences our brains and our behavior. Everyone has anecdotal observations from tailgates or weddings, but such observations do not hold up to the scientific rigor of controlled, experimental studies in the laboratory. Crayfish, like humans, are social animals, which allows them to serve as a model for understanding the neurological and social effects of alcohol. Once Herberholz and his students knew that crayfish responded to alcohol similarly to humans, they could start to figure out exactly how alcohol consumption influenced crayfishes' social interactions and what was going on in their nervous system.

Using animals to research alcohol's effect on social behavior has a long history. Historic studies with alcohol and nonhuman primates led to poor insights into human behavior due to the complexity of monkeys' and apes' social structures and nervous systems. But with crayfish, that's not an issue, making them an unorthodox but perfect animal to figure out what happens to the nervous system when you mix alcohol and socialization. To figure out how alcohol consumption influenced social behavior, Herberholz and his students took juvenile Red Swamp Crayfish and assigned them to one of two groups (there is no legal drinking age for crayfish, so no laws were broken here). Crayfish in one group were raised in isolation and never experienced another crayfish (the isolated treatment). The second group was raised in a community tank, with 50–100 other crayfish (the communal treatment). Crayfish raised in the communal group were socialized with other crayfish, constantly fighting and communicating with each other to settle disputes over food and shelter.

Then, crayfish from both the isolated and communal treatments were exposed to different concentrations of alcohol. After exposure, crayfish in both treatments exhibited the typical drunk behaviors described above: aggressive-like postures, spontaneous tail flips, and eventually the trademark carapace-down, legs-up position of a drunk crayfish. But what was amazing is how the behaviors of the social crayfish differed from those of the isolated, loner crayfish. Crayfish from the communal group would start exhibiting the effects of alcohol much faster than the isolated crayfish, meaning social crayfish get drunk faster. Crayfish that were isolated took, on average, twenty-eight minutes to start exhibiting the spontaneous alcohol-induced tail flips, but crayfish that had been socially housed started exhibiting this behavior in only twenty minutes.

Furthermore, Herberholz and his students found that they could isolate the control of the intoxication behaviors down to the individual neuron, which are nerve cells vital in our nervous system's ability to send electrical and chemical messages throughout the body. This meant that they could figure out the exact neural mechanisms that were responsible for the crayfish getting tipsy or drunk at different rates. When crayfish were surrounded by other crayfish, alcohol lowered the threshold of the neuron's response, which decreased the amount of time it took for them to exhibit drunkenness.

The results of this work were published in the *Journal of Experimental Biology* and attracted several online media outlets that covered "getting crayfish drunk for science." Although speculative, the results of this study can give insights into the use and abuse of alcohol in humans. The reduced sensitivity to alcohol that was observed in the socially isolated crayfish may be the same underlying mechanism that relates to increased alcohol consumption in socially isolated humans. This study also demonstrates a conserved effect of alcohol on movement across animals, which means that other impacts of drug use such as withdrawal and habituation may be worth investigating in crayfish.

Getting crayfish drunk, high, and spaced out in the laboratory is both a comical and informative method of learning about drugs and how they affect nervous systems and animal behavior. Not to mention it makes for some interesting dinnertime stories or happy-hour conversations. But what about crayfish in nature? Do crayfish downstream from the local dive bar ever get drunk? Or can burrowing crayfish get a contact high if the smoke from a nearby rock concert sneaks into their chimney? It turns out that although those two scenarios are unlikely, crayfish in the current day and near future may be more mellow than ever thanks to a healthy dose of antidepressants.

## Crayfish, COVID, and Crises

From the 2010s into the 2020s, the number of antidepressant prescriptions increased tenfold. The COVID-19 pandemic added to these numbers, with a further threefold increase in antidepressant use. Recent data suggests that one in every eight individuals in the United States is prescribed an antidepressant. Being the researcher I am, it's tempting for me to think, "Humans are going to solve this mental health crisis . . . by learning about and studying crayfish!" My own enthusiasm for these creatures aside, crayfish actually may be able to help our understanding of how antidepressants affect an animal's brain and behavior. On top of that, crayfish can be used to study the ecological consequences of the rise in antidepressant use. Right now, somewhere under a rock or a log there are crayfish feeling so relaxed and antidepressed that it may be a detriment to their survival. The waters that crayfish, fish, and other aquatic organisms inhabit are steadily becoming a chemical slurry of pharmaceutical pollutants.

Increased use of pharmaceutical drugs means that more of these chemicals are present in the environment, making their way into waterways through human urine and feces. Antidepressants can last for well over a year floating down rivers and streams, eventually settling in the sediment. In addition to chemicals passing through us, there is also the possibility of large-scale pharmaceutical pollution due to leaky wastewater systems or chemical dumps from pharmaceutical factories. Now, most of the time, the concentration of drugs in an aquatic ecosystem is so minuscule that there is no direct impact on humans. No human is going to dive into an antidepressant-polluted lake and come out feeling stress-free. But the same cannot be said for other, smaller animals, like crayfish, because drugs may quickly accumulate in an ecosystem's food web through biomagnification.

In a drug-polluted ecosystem, stream-dwelling invertebrates like crayfish accumulate trace amounts of the drugs by consuming substrate or plant matter as well as passively absorbing them through their gills. And climbing up the food web, larger crayfish then consume many of the smaller invertebrates, magnifying the levels of drugs in their system even further. Top aquatic predators like bass or even terrestrial predators like raccoons will consume a few dozen crayfish, thus biomagnifying the drugs even further—essentially making a crayfish snack into a hit of a random assortment of drugs.

Through biomagnification, chemicals can accumulate in an environment and add up to a dose almost as potent as a human's dose. Even some of the most obscure animals, like platypuses, have been found to contain bio-

82

magnified doses of antidepressants that are half of a typical human dose, even though the average platypus weighs less than five pounds and the average human weighs 150 pounds.

With rising antidepressant use and pharmaceutical pollution, researchers came to the crayfish to learn what effect dumping drugs into the environment has on the organisms within. A study conducted by Alex and Lindsey Reisinger (a married duo with expertise in biogeochemistry and community ecology, respectively) and their collaborators sheds light on the issues highlighted above.

Instead of conducting their studies in a featureless glass aquarium in a temperature-controlled room with blaring lights overhead, this team worked with larger, naturalistic enclosures placed outside, called mesocosms, which are outfitted with natural shelters and vegetation. Mesocosms are ideal for ecological studies, as the environment closely mimics nature, with animals in outdoor mesocosm experiments experiencing natural fluctuations in temperature, precipitation, and other variables. In this case, the researchers collected Spinycheek Crayfish (*Faxonius limosus*) from a stream in New York state, where these studies were conducted.

In this area full of mesocosms, crayfish were placed in naturalistic enclosures and drugs were added, allowing the researchers to understand exactly how pharmaceutical pollution alters crayfish behavior—and potentially their entire ecosystem. The researchers' drug of choice was citalopram,

A Spinycheek Crayfish (*Faxonius limosus*). Wikimedia Commons.

commonly sold as Celexa. Citalopram belongs to a class of antidepressants known as selective serotonin reuptake inhibitors (SSRIs), which help regulate your body's serotonin levels, ultimately balancing irregular neurochemistry, heightening mood and happiness. With 10 million to 36 million adults in the United States being prescribed SSRIs such as citalopram annually, a lot of these chemicals get washed into waterways and absorbed by the bodies of other organisms. Understanding the consequences of these chemicals is a new focus for many biologists, and crayfish are an ideal candidate to explore their impact.

In the Reisingers' simple experimental design, half of the crayfish in a mesocosm were not given any drugs, and half were dosed with 500 nanograms of citalopram in the water. The dosage was based on data showing that natural bodies of water contain roughly this amount. Using environmentally realistic concentrations makes the results of the Reisingers' study directly relatable to natural (but polluted) ecosystems, not just naturalistic enclosures. A standard human dose is twenty milligrams, which is 40,000 times more than the dosage used in this study; but remember, crayfish are small, and this makes them lightweights.

After the researchers put crayfish in their respective enclosures, either with or without citalopram, they recorded their behavior over the following weeks. They gathered crayfish from both treatments (control and citalopram) and gave them several ecologically relevant tasks outside of their initial mesocosm. Several results from these trials are worth highlighting, because it turns out that crayfish respond to antidepressants similarly to humans.

In one test, crayfish from both treatment groups were tested on their willingness to wander out in the open and gather food. The control group responded appropriately and hesitantly, as any crayfish should. Once sighted by a predator (even though there were no predators in this experiment), a crayfish caught out in the open may be gobbled up in an instant. Control crayfish protected themselves by staying underneath a shelter for the majority of the time. But for the cool, calm, and collected citalopram crayfishes, predators were the least of their worries. Crayfish who had been given the antidepressant ran out of their shelter and went straight for the food— spending 400% more time where the food was than the control crayfish did.

In a second set of trials, when exposed to the scent of a predatory fish, crayfish given the antidepressant were more likely to forage and spend time outside of the safety of their shelter. In contrast, crayfish in the control treatment behaved appropriately, sheltering themselves, knowing that danger was imminent. In all cases, when the environment contained small but realistic concentrations of antidepressants, a crayfish's willingness to engage in risk-taking behavior skyrocketed.

To a human, SSRIs and other antidepressants can make previously anxiety-inducing scenarios become mundane. Antidepressants may give that small boost of serotonin needed to become comfortable with the tasks of modern human life. But for a crayfish, the same antidepressants override hundreds of millions of years of evolution that has made these animals wary and calculated.

A world filled with relaxed, risk-taking, fearless crayfish may not seem like a big deal, but extrapolate these results to the scale of an entire ecosystem, and things get out of hand quickly. Instead of a single crayfish in a mesocosm, imagine a stream that may contain hundreds, if not thousands, of crayfish. These animals are typically waiting until the sun sets before they leave the safety of their burrows. But the results of the Reisingers' study (as well as those of several other researchers who have independently found similar results) suggest that, though at a small scale, a community of unafraid crayfish full of antidepressants can have big impacts on the ecosystem.

Crayfish with reduced predator responses means more food for the bass, trout, catfish, herons, raccoons, and every other organism out there looking for a shellfish dinner. This cycle may start out good for the consumers of crayfish. But over time, this overconsumption could throw off the environmental teeter-totter that keeps ecosystems in balance. And with more crayfish being consumed, this means fewer crayfish to break down leaf litter, fewer crayfish to consume small invertebrates, and fewer crayfish to create burrows that other animals desperately rely on. Everything quickly gets out of whack.

The effect of biomagnification is also worrisome, because if the crayfish are accumulating drugs and the crayfish play an integral role in the environment's food webs, then that means all the larger predators will also be accumulating the drugs. This biomagnification may impact not only aquatic predators but also terrestrial ones that consume crayfish. Due to the way ecosystems are balanced, if you mess with one species, you end up messing with them all.

Pharmaceutical pollution is becoming a bigger issue year after year. The research being conducted on crayfish and other organisms warns of a future where drugs flow freely throughout aquatic ecosystems, changing not only the behavior of animals but potentially the entire way the environment operates.

## Through Crayfish, the Whole World

Over 150 years ago, Huxley introduced the "insignificant" crayfish to the broader biological community. His hope was that crayfish would be used to resolve some of the most difficult problems in biology. Nowadays, biologists recognize that these creatures are far from insignificant, and Huxley's dream became reality. Crayfish are being studied in all realms of biology; they are being studied in the wild, in makeshift laboratories, and even in some of the top medical institutes in the world. Researchers use them to figure out what happens to a crayfish's brain when it gets drunk or to see how increases in pharmaceutical pollution may impact the future of freshwater ecosystems. Some of the largest biological advancements have been brought about thanks to crayfish, and there are surely more discoveries on the horizon for these laboratory wonders.

CHAPTER 5

# Communicating
# with Urine

## The Umwelt

ONE OF THE CARDINAL SINS of studying animal behavior is anthropomorphism—the attribution of humanlike characteristics to a nonhuman animal. Pet owners are serial anthropomorphizers, as they interpret their pet's behaviors through the lens of their own world and not the pet's world. If a dog has its mouth open and is exposing its teeth, an owner might assume it is "happy and smiling." If a dog pees in the corner and then runs away from its owner, it's "feeling guilty." Although a bit of anthropomorphism is fine for casual conversation or storytelling (as you'll see throughout this book), it can be dangerous and uninformative to biologists. When humans attribute human characteristics to nonhuman animals, they assume that other animals experience their environment in the same way that humans do. This is rarely the case.

It is easy to forget that cats will urinate anywhere and everywhere if they become stressed, or that dogs' greetings don't involve a physical handshake but rather a chemical handshake of sniffing each other's behinds (where the anal glands are located). With only a few sniffs, dogs can learn about the sex, hormone levels, and physical condition of another dog. All the chemical information in a dog's anal gland is widely available to other dogs thanks to their acute sense of smell, a capability completely inaccessible to humans.

On top of this, the sensory capabilities of most organisms differ from humans'. Armed with different visual receptors and neural physiology, dogs are dichromatic and have color receptors to see only muted blues and yellows. In another example, flowers don ultraviolet colors that are imperceptible to us but are like a highlighter for flying insects, drawing them in like a shiny target for pollination. Vision aside, strange animals called knife fish communicate through electrical pulses, and doves and other birds use the earth's magnetic fields like a compass for navigation. Bridging this sensory gap is key to understanding animal behavior.

EARLY IN MY TRAINING as a behavioral biologist, I learned the traditional way of getting into the sensory world of other animals by entering their umwelt [oom-velt]—a German word that roughly translates to "environment" or "surroundings." The umwelt was popularized by Baltic German biologist Jakob Johann von Uexküll, who believed that much could be learned about animal behavior by imagining what life is like without the human sensory tool kit and instead imagining the sensory world of other animals. Entering the crayfish umwelt means understanding what it's like to have long antennae, stalked compound eyes, and hundreds of chemically sensitive hairs covering your body. Von Uexküll's umwelt acknowledges that although two animals can experience the same environment, the way in which they perceive and use this environment varies.

I and other crayfish obsessives spend restless nights trying to understand how and why crayfish interact with the environment and one another in the way that they do. Getting into the umwelt of a crayfish is a necessary part of understanding its biology, which requires a peek into the inner workings of a crayfish's sensory tool kit. Occupying the umwelt of crayfishes is one of the only ways to truly appreciate their diversity, understand their behavior, and even potentially help protect these organisms. This requires forgetting the dominant human senses and replacing them with those of a crayfish. It also requires being open-minded and creative and, maybe most important, learning a lot about urine.

# Life Underwater

After doing a cannonball into the water, most people will immediately shoot back up to the surface, gasping for air. For those seconds underwater, everything is different. Sounds are blurred, vision is reduced, and other senses, like smell, are completely disabled. Getting into the umwelt of a crayfish requires first recognizing the differences between land and water, because most crayfish live their entire lives underwater—which may as well be a different planet to humans.

Our journey into the crayfish umwelt will explore the sensory world of crayfishes through the lens of a typical tertiary burrowing crayfish, which spends its entire life underwater. Although some crayfishes, like the primary burrowers, spend plenty of time on land, much of their sensory biology is unexplored and is assumed to be similar to that of a tertiary burrowing crayfish. By seeking to experience the crayfish umwelt, we'll go through what senses a crayfish uses and how its experiences and interactions differ from that of land-dwelling bipeds such as us.

Light, sound, and chemicals all interact with liquid water differently than with gaseous air. Thus, aquatic organisms like crayfish often use different senses than land animals. Settling a dispute or giving directions is hard enough on land—but even a basic task becomes complex underwater. Imagine being lost and trying to get directions home, but instead of clear air and stable ground, you are surrounded by flowing, murky water with a constantly changing landscape of rocks, mud, sticks, and predators. This is the reality for crayfishes; they need to complete the basic tasks of animal life—locate food, not get eaten, find mates, and communicate with each other—all underwater.

Although some crayfishes live in clear water, most species live in muddy, sandy, turbid water with less than one foot of visibility. And because most of the top crayfish predators are visually oriented daytime hunters, crayfish are primarily active in darkness—leaving their shelters only when the coast is clear and predators aren't easily able to hunt. If you are primarily active at night and spend the day in a low-visibility environment, having good vision isn't very useful.

As a general rule, the size of an animal's sensory organs relative to its overall body size can help you determine how important that sense is to that animal. Elephants have massive ears that help them hear sounds nearly a mile away, and dragonfly heads are almost entirely eyes, which allows them to intercept flying prey with ease. Applying this rule to crayfish vision, it

An up-close view of the compound eye of a crayfish. Each eye is composed of hundreds of individual units called ommatidia. Photo by Mark Harris.

becomes even clearer (or blurrier, to be more visually accurate) that vision has taken a back seat to other senses.

As if they were conjured up in a science fiction novel, crayfish have a pair of stalked, beady little eyes that can rotate nearly a full 360 degrees. Unlike the complex camera-like eyes of humans, crayfish have compound eyes, which means that they are composed of hundreds of smaller visual structures called ommatidia that contain everything that makes an eye an eye: a cornea, a lens, and photoreceptor cells to determine brightness and color. Compound eyes are bad at forming clear images or detecting faraway objects; a crayfish would miserably fail at distinguishing between the letters JPYLA and LPYJA at its yearly eye appointment. Contrast the eyes of a crayfish with the eyes of humans and many other vertebrates, which can create crystal-clear images across long distances. But what crayfish lack in image formation, they make up for in their ability to detect movement—an invaluable ability for an organism that is eaten by nearly every predator around.

Anyone who has attempted to hand-catch a crayfish is familiar with their ability to quickly detect a hand as a threat and escape—making even a skilled crayfish biologist look clumsy as they chase one around in the water. A crayfish may be unable to distinguish between a raccoon hand reaching into the water, a bass swimming nearby, or a heron flying overhead, but what it can do is quickly recognize motion and escape to live another day. With poor eyesight, crayfish do not rely heavily on vision in their daily life. Laboratory experiments performed over the last few decades confirm this, as crayfish can find food, mate, and even fight when blindfolded.

Animals with good eyesight develop color vision to further enhance their ability to find and locate objects. Color vision in humans and other animals has evolved to detect uniquely colored objects, like yellow dandelions or red apples. But being active at night and in murky waters doesn't allow for efficient visual detection, so color vision does not need to be well developed in crayfish. Although few researchers have delved into the world of cray-fish color vision, it is known that some crayfish species have two classes of photoreceptors, sensitive to red and blue wavelengths—meaning that crayfishes should be dichromatic, seeing two different colors. But behavioral evidence that demonstrates color vision is lacking.

This completes the first step of our crustacean transformation and journey to embody the umwelt of a crayfish. Our complex camera eyes have been swapped for rotating, beady eyestalks that work more like motion detectors. The next step involves a sense that is important for humans yet nonexistent in crayfish: hearing.

WHEN WE'RE TUNED IN, the sheer number of sounds that occur on land can be overwhelming; air is a perfect medium to carry the vibrations that human ears recognize as sounds. Birds, amphibians, and mammals all use the physics of air to communicate over long distances, whether it's with a trill, a song, or a roar. Animal sounds can be so abundant that when a lull in noise occurs, human brains become instilled with fear. But in water, making noise is less common and is most notably reserved for some of the largest underwater organisms, like dolphins and whales.

Based on our current understanding, crayfish are deaf. They have no sensory organs capable of hearing. But just because they have no organs with which to hear noises, it doesn't mean they are unable to make noises. When picked up out of the water, many species grind their mouthparts together to create a scraping, squealing noise. Since this sound is produced only when a crayfish is out of water, it is hypothesized that the noise is created to scare off or confuse predators, potentially allowing the crayfish to use its claws to deal some real damage.

Two other instances of crayfish sound production have also been reported. When taken out of the water, the large Australian Murray River Crayfish (*Euastacus armatus*) flexes its abdomen to create a sound similar to the mouth-grinding sound that other species produce. The now extinct Tasmanian tiger used to prowl the areas that the Murray River Crayfish inhabits, leading to speculation that crayfish may have used its abdomen-grinding noise to ward off that wolflike marsupial predator. There are also reports of a burrowing crayfish species, the Prairie Crayfish (*Procambarus gracilis*), using mouth-grinding noises during social interactions on land. Whether these crayfish use these sounds to communicate with other crayfish or predators is unknown.

With both vision and hearing reduced or nonexistent in the umwelt of a crayfish, our journey toward understanding crayfish senses is progressing. Nonetheless, crayfish still need to be able to interact with the environment and navigate their surroundings. The way they accomplish this is foreign to humans. Here, creativity becomes essential, as one departs from the everyday senses and begins to learn what it truly means to be a crayfish.

BLINDFOLDING A CRAYFISH was a favorite pastime of nineteenth-century biologists. Harvard Medical School doctor George V. N. Dearborn was among the first to describe the behavior of blindfolded crayfish in his whimsical article "Notes on the Psychophysiology of the Crayfish." Dearborn was also the first person to use inkblot tests in human psychiatry, well before the more well-known Swiss biologist Hermann Rorschach popularized this method with the Rorschach test. Dearborn used "tin helmets" to blindfold crayfish, but modern studies just place opaque tape over their eyes, plop them into an aquarium, and watch what they do. Even when blinded, crayfish can find their way around with no issue at all. For a crayfish, being unable to see is the norm—they spend most of their lives tucked away under rocks or in burrows in the dark. But what makes crayfish so comfortable navigating without vision? What prevents them from running carapace-first into a rock?

Compare the way a blindfolded crayfish navigates its surroundings with the way a human navigates with their eyes closed. Imagine you're blindfolded in an unfamiliar room filled with everyday objects: a table, a bookshelf, a couch, and other household items. Navigating this room with your eyes covered is a nightmare, as you must creep around using your hands and feet as probes to detect the objects in the room. When you encounter an object with one of your appendages, a single touch can tell you the object's temperature, texture, and even moisture. With a series of touches, you eventually form a mental picture of your surroundings—you can feel the height and depth of the dinner table and remember its location in the center of the room. After a lot of probing, your mind starts to form a visual image of the room you've explored and you can navigate this area with some semblance of confidence, despite being blinded.

Now imagine you're a blindfolded crayfish (with your choice of contemporary opaque tape or a vintage tin helmet). When blindfolded and plopped into an aquarium, your exploration mode instantly kicks in. Instead of probing with your arms and legs, flailing like in a ritualistic dance, you would be comfortable with this type of task because of your pair of long whiplike antennae that shoot out from the front of your body. Antennae function as two extra-long arms that can touch, feel, and explore the environment—providing information about the surroundings. In some crayfish species, the antennae are longer than the body—this reach gives the crayfish an extraordinary ability to learn about a new environment.

While exploring, crayfish wave their antennae side to side to scan their environment. Each antenna moves independently, which allows for sweeping coverage of the area in front of the crayfish. When an object is detected, the crayfish moves closer, with its antennae still in contact, until it can get up close and personal to "feel up" this new object before continuing to explore with its antennae.

As the antennae meet an object, the crayfish is flooded with information, because a single crayfish antenna has around 7,000 hairlike touch receptors, called setae [see-tee]. Every time the antennae touch an object, a response from hundreds of setae sends a neurological signal to the crayfish's brain, which provides information on the size, shape, texture, and temperature of the object. The sensory hairs and gangly antennae extend the tactile capabilities of the crayfish much like how having two highly sensitive foam pool noodles dangling from your forehead would surely help in detecting and navigating the world around you. (In all seriousness, you would get used to it.)

In the same way that humans form an image of their surroundings while blindfolded, experimental evidence by comparative neurobiologist David Sandeman demonstrates that crayfish may be capable of a similar feat—all thanks to the sensitive hairlike receptors covering their bodies and their antennae. The possibility of crayfish producing mental maps is intriguing and would be beneficial to them in navigating their territories or familiarizing themselves with a new environment.

Although the antennae are the hot spot for touch-sensitive hairs on the crayfish, they are not the only place that crayfish can feel when touched. Humans have heightened sensitivity on our fingers and lips, but our skin, from our head to our toes, is also capable of sensing touch. In the same way, crayfishes have the greatest density of sensory hairs on their antennae, but

A Hillbilly Hairy Crayfish (*Cambarus polypilosus*). The scientific name for this species is derived from the Greek words *poly*, for "many," and *pilose*, for "hairs," which relates to their fuzzy appearance. These hairs cover their body and presumably help this species navigate the rocky substrate in which they burrow in the various Tennessee streams they inhabit. Photo by Zackary A. Graham.

these receptors are also spread throughout their exoskeleton—all the way from their claws down to their walking legs and even their abdomen.

When these hairs are viewed under fancy scanning electron microscopes (SEMs), several forms of these hairlike structures can be observed, including ones that look like normal hairs, ones with intricate feather-like structures, and others shaped like an upside-down question mark.

These hairs are sensitive not only to physical touch but also to hydrostatic movement. This means that crayfish can detect the movement and direction of flowing water—just like humans can tell whether a cool draft of air is blowing on their toes or on their neck.

If a catfish swims nearby and creates a current of water from its movement, the crayfish will detect this current and tail-flip away without even having to be face-to-face (or more realistically, face-to-claws) with the fish. If a predatory aquatic bird like an egret lunges at a wandering crayfish from above, the crayfish can detect the disruption of the water current and allow for a prompt escape. The sensory hairs give the crayfish the ability to acquire and respond to information from all angles.

# CRAYFISH ANTENNAE

Crayfish antennae are composed of up to 200 individual segments, which allow for whiplike flexibility. The segments are also thinner toward the tip of the antennae, which ensures that the antennae do not get jammed or stuck while exploring.

The vast majority of crayfish species are equipped with a standard-issue set of antennae. But some species have unique or exaggerated features on their antennae. Cave crayfishes are known for their unique antennae, which are often much longer and thinner than those of a typical stream- or river-dwelling crayfish. Living in complete darkness, cave crayfishes have no sense of vision. Most cave crayfishes have traded their loss of eyes for an increase in the length of their antennae. In theory, because blind cave crayfish are "blindfolded" and have lost their vision, their extra-long antennae are an adaptation that helps them navigate in darkness. Although not all cave-adapted crayfishes

A juvenile crayfish with spread antennae exploring its environment.
Photo by John Abbott/Abbott Nature Photography.

The Spider Cave Crayfish (*Troglocambarus maclanei*), a cave species from Florida with the longest known antennae of any crayfish. The antennae extend out of the frame of this photograph. Photo by Chris Burney.

are blind, elongated antennae occur in all four North American genera that have cave species: *Troglocambarus, Orconectes, Procambarus,* and *Cambarus.* One cave species, the Spider Cave Crayfish (*Troglocambarus maclanei*), native to Florida, holds the record for the longest antennae, which reach on average 2.8 times its body length.

With another update to the crayfish umwelt, the scenario now includes deafness, hazy vision, and a heightened sense of touch and navigation. Although these senses are well suited for a solitary lifestyle, it's important to note that crayfish are often the most abundant invertebrates in aquatic environments. They spend much of their lives alone yet are inherently social animals, whether by choice or by necessity—true introverted extroverts. Understanding how crayfish communicate with neighbors, mates, offspring, and even rivals is essential. Remarkably, this communication is achieved by spraying urine directly toward the face of another crayfish—a behavior that invites, but perhaps resists, comparison to human interactions.

# Communicating with Urine

From the perspective of the human umwelt, other animals defy sensory logic. In complete darkness, male moths can somehow navigate through a forest full of trees—only to stop at the one tree that has a female. Cave-adapted animals without eyes, whether fish, salamanders, or crayfish, can detect and escape nearby predators. How are these behaviors possible? How can these animals make complex decisions like detecting a mate or fleeing from a predator without any sophisticated visual, auditory, or tactile senses to feed them information?

*Sniff, sniff.* You might have smelled this from a mile away, but the answer relies on these organisms' sense of smell—more specifically, their ability to release, detect, and respond to chemicals in their environment. To clarify the behaviors presented above, female moths release a pheromone that acts as a lure to draw in the male. And crayfishes also have an acute sense of smell they can use to detect predators, mate with other crayfishes, and even communicate the outcome of a fight—all of which can be done without vision.

What humans recognize as a sense of smell is nothing more than the ability to detect chemicals in the environment, in the same way that vision is the ability to detect light in the environment. Some chemical compounds, like air, smell like nothing. But add in some methane, and you will miss the smell of nothingness. Throughout this section, I use the terms "smelling" and "chemical sensing" interchangeably, because what humans call a sense of smell (and taste) is technically our ability to sense chemical compounds.

Biologists' understanding of the chemical world of animals lags twenty to thirty years behind other fields in sensory biology because of how difficult chemicals are to study and identify. Humans are biased. For us, it is easier to understand the coloration of a cardinal than the invisible release of chemicals by an ant. Chemicals and smells influence human lives; they do so in our ability to detect good scents (this apple smells amazing) and bad scents (this apple smells rotten), which have enabled us to identify which foods are nutritious and which will make us sick.

Crayfish live in a world where sending and receiving smells is crucial. They need to send out scents, sniff scents from others, and then decipher what information is being conveyed, all with chemicals that are transmitted through a liquid environment.

EARLY BEHAVIORAL STUDIES with crayfish make it clear that there is something going on in the umwelt of the crayfish that is distinctly different from the human umwelt. A simple experiment reveals how invisible chemi-

# A URINE RÉSUMÉ

Why urine? What is so special about this substance that makes it such a useful form of communication in crayfish? Oddly enough, communication through urine is common in the animal kingdom. Think of how many dogs have urinated on a single telephone pole in your neighborhood. It turns out that you can learn a lot about an animal based on what their urine smells like.

An animal's urine is like its résumé, just in chemical form. Because urine naturally concentrates the waste products in your body, urine composition gives you information about the past experiences and achievements of that animal—just as a résumé would. If you are stressed, whether you are a crayfish, dog, or human, you'll create waste from the production of stress hormones, which accumulate in urine. Humans even use urine samples to determine prior drug use or potential infections. In crayfish, information regarding sex, aggression, prior fighting history, and species are all conveyed through urine. Therefore, by sniffing the urine of a fellow crayfish, you'll know exactly what to expect.

---

cals guide a crayfish's behavior. Take a cup of water from a female crayfish's aquarium and pour it into a male's aquarium. Out of nowhere, the male will explore as if he were on a mad hunt for the female, despite being alone in his aquarium. If you really want to play tricks on the poor male, add in a rock from that same female's aquarium, and the male may attempt to mate with that rock!

Turn the tables and pour water from a male's aquarium into another male's aquarium and the resident male will be on high alert—raising his posture to make himself look large and displaying and opening his claws—ready to fight at any moment despite being completely alone. Because the crayfishes being studied in these scenarios were isolated, early biologists realized that these responses only make sense if the crayfish can sense (undetectable to us) chemicals in the water.

Some chemically minded biologists wanted to investigate what the chemical substance was that these crayfish were responding to. Was it a chemical being put out by their exoskeleton, like a crayfish version of body odor? Or was it a scent specifically produced by each sex with specialized glands?

None of these early ideas panned out, and when the biologists did discover what the crayfish were sending out into the environment, they were probably slightly disgusted: These animals communicate with urine. The hints have been building throughout the chapter, but now it's time to experience the scent of urine just as a crayfish would.

A closer look at crayfish anatomy makes it clear that they are efficient urine-spraying machines. Instead of expelling urine from their excretory system as waste like most animals, crayfishes have evolved the ability to store their urine in their bladder over long periods—tactfully opening the floodgates when necessary. A 1980s study estimated that around 3% of a crayfish's total body weight is urine. In my 200-pound body, that would equate to enough urine to fill one and a half two-liter soda bottles, or ninety-six ounces total!

Crayfish urine is a prized possession that doesn't go to waste. To ensure that the correct amount of urine is released, crayfishes have a pair of circular structures called nephropores tucked just beneath their eyes. The nephropores are connected internally to the urine-storing bladder. Nephropores are tiny openings controlled by sphincter muscles, so crayfish have fine-scale control of when and how much urine is released by modulating how open or closed the nephropores are. In isolation, crayfish will occasionally crack open their nephropores and leak out some urine just to produce a general scent that makes any other crayfish aware of their presence. But thanks to an ingenious set of experiments, astacologists now know that in other scenarios, urine release is on full blast.

HAVING A STEADY HAND is an important gift in the medical field, where pinpoint accuracy is required for surgery or injections. Thomas Breithaupt, a professor at the University of Hull in England, was aware of this when he assigned his student Petra Eger to a task that requires surgeon-like accuracy. Breithaupt is a sensory ecologist who studies how animals use their senses to interact with other animals and their environment. Throughout his career, he has studied several aquatic animals, including catfish and lobsters, but among his lengthy list of publications, as he told me in my personal conversations with him, crayfish, and the chemicals they release, are his favorite.

Breithaupt knew that chemical communication was a dominant force in the lives of crustaceans. But because crayfish are aquatic animals, their urine is not like ours. Human urine turns yellow and can be seen in water when we are dehydrated; crayfish urine is always crystal clear. Breithaupt thought that if he could come up with a way to visualize the urine release, he could

An illustration of two crayfish excreting urine from their nephropores during a social interaction. Illustration by Liz Pavlovic.

study when it was released, how much was released, and in which directions it was spread around—three questions that tie directly into the social and sexual behavior of crayfishes.

This is where Eger came in. She was tasked with finding a way to visualize the chemical release of urine in crayfishes. Eger would have to conduct dozens of crayfish surgeries for this project. Lucky for the crayfish, she had surgical finesse from giving hundreds of injections in her days as a nurse. Breithaupt and Eger figured that because the urine was released as a liquid, they could use a fluorescent dye to visualize it, much the same way that food dye droplets are briefly visible in water before they become diluted. Introducing the fluorescent dye requires a precise injection of fluorescein (a common fluorescent medical-grade dye) into the crayfish's pericardium, or the tissue that surrounds the heart. In theory, since the pericardium is connected to the bladder, a fluorescein injection into this area should lead to a crayfish spewing out a plume of fluorescent green pee.

Getting the fluorescein into the pericardium is no easy task. Inject a needle too shallowly and never reach the pericardium—no fluorescent green pee. Inject the needle too deep, and you risk injuring the crayfish—and again, no fluorescent green pee. On top of the pinpoint injection accuracy required, immediately after removing the needle, you need to fill the hole in the exoskeleton by sealing it with beeswax to ensure that no hemolymph (the equivalent of blood for invertebrates) is lost. But if you get the injection just right—not too deep but not too shallow—what occurs should be considered for the eighth wonder of the world.

When the fluorescein dye makes its way into the bladder, the dye mixes with the urine and allows for the visualization of urine release in crayfish. Because urine is used during social interactions between crayfish, if you put two dye-injected crayfish in an aquarium, you will see each crayfish's nephropores open and a frontal cone of lime-green urine blast out in front of the crayfish straight into another crayfish's face. The first time he and Eger produced a successful injection, and thus could see the urine, Breithaupt started to think of the endless experiments he could conduct now that he could visualize urine release. Before their experiments, it was unknown exactly how much or when urine was released. But thanks to the fluorescent visualizing of crayfish urine, Breithaupt and Eger could now learn about the chemical world of crayfishes in a way that no one else had.

RELEASING URINE is one thing, but it needs to get out into the world so other crayfish can get a sniff. Imagine if an aerosol can of pepper spray had no propulsion mechanism; everything would just dribble down the can.

Based on the speed and rate at which he saw the urine released, Breithaupt knew that something had to be propelling it, which was allowing the urine to form a cone shape in front of each crayfish. Conversations I have had with Breithaupt demonstrate his extreme familiarity with crayfish anatomy, as he exclusively uses the scientific names for structures, whereas many prefer the common names. Even for me, keeping up with this language can be difficult. This familiarity with the crayfish's body gives Breithaupt a strong understanding of the animals he studies. Thanks to several experiments, he figured out that the prominent propulsion mechanism comes straight from the crayfish's internal gills. As urine is released from the nephropores, crayfish can create a strong current by beating their gills, which propels the urine forward—ultimately creating that cone-shaped urine blast that he was witnessing.

All this urine sending is interesting and entertaining, but it is useless unless there is a crayfish on the receiving end. It is no coincidence that Breithaupt was noticing crayfish spray urine when two crayfish were face-to-face. That's because nestled between the two long walking-stick-like antennae resides a smaller pair of antennae, called antennules, which are the nose of the crayfish.

The antennules are a fraction of the length of the antennae yet are jam-packed with setae, all of which are highly sensitive to any environmental chemical in the water. In isolation, a crayfish's antennules remain still. But during exploration or when a new odor is detected, the antennules are flicked so rapidly they appear as a blur. Antennule flicking is akin to a human sniff—these behaviors increase the chances of coming in contact with chemicals that could provide valuable information about one's surroundings.

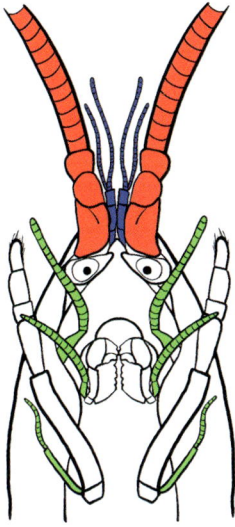

An illustration of crayfish chemical sensory organs. Antennae are in red, antennules are in blue, and fan organs are in green. Illustration by Hannah N. Holbert.

The crayfish's antennule flicking is paired with the movement of three pairs of appendages outside of its mouth, which Breithaupt describes as fan organs. The fan organs are little flexible tendrils that are designed to fan water from all directions into the crayfish's antennules to maximize contact with urine in the water. The fan organs can also double as a method to spread out and redirect the urine the crayfish releases from its nephropores to ensure that urine spreads.

In addition to the antennules and fan organs, crayfishes have chemical-sensing hairs packed throughout their exoskeleton, with an especially high density of sensors on their claws. Although these hairs are similar in appearance to the mechanosensory touch hairs on their body, they are neurologically sensitive to smells, not touch. Humans' noses are fairly one-dimensional; the scent needs to come from the front and then reach the nose. But crayfish can smell a scent from all directions, whether it comes from behind them or straight in front of them.

Breithaupt sees urine-borne communication by crayfish as an ingenious adaptation to aquatic life. In a pool with little to no water movement, traces of chemicals can be tracked for hours and even days after they are emitted. Further, since chemicals last longer in water, this can allow them to travel in flowing trails to be communicated to others downstream. Chemical communication can also occur in pitch-black or completely murky waters, making this the perfect method of sending out information in the environments where crayfish thrive. Above all, chemicals are king in crayfish.

# MORE HAIRY CRAYFISH

A typical crayfish claw is filled with dozens of small dimples that are packed with setae, some of which are sensitive to touch, whereas others are sensitive to chemicals.

In fact, some species of crayfish, like the Jackson Prairie Crayfish (*Procambarus barbiger*) and *Procambarus llamasi*, have dense clumps of setae that are often located on the outside edge of their claws, which gives the appearance of their claws having mittens, because the setae are so dense. Why some species have these large patches of hairs has yet to be investigated, but they likely play a role in chemical communication.

A Jackson Prairie Crayfish (*Procambarus barbiger*). Photo by Guenter Schuster.

Paul Moore, professor of biology at Bowling Green State University in Ohio, and longtime sensory biologist, describes the chemical world of a crayfish in a stream with a unique metaphor: "Imagine you are at a club with a disco ball, with shimmers of light flying everywhere, except instead of light, it's chemicals." Bombarded from all angles, you smell your neighbor two rocks down whom you have been in a turf war with for weeks, the same group of females across the way, and dozens of other local urine senders. This is what it's like to be a crayfish.

With vision blurred, hearing absent, touch-sensitive antennae, sensory hairs scattered across the body, and a keen chemical sense, the necessary parts of the crayfish umwelt have been assembled. These components provide the foundation for understanding the complexities of crayfish social behavior. Being blindfolded, deaf, and guided almost entirely by touch and chemical cues creates a recipe for uniquely structured social interactions—and indeed, the result is as unusual as expected.

## Battle of the Benthic

As I tower over a dozen elementary-school-aged kids crowded around me at about elbow height, I point out the walking legs, long antennae, large abdomen, tiny eyes, and robust claws at the front end of the crayfish body. Microscopes, test tubes, and other fancy lab equipment surround us, but the animal in my hand is the center of the kids' attention. The bolder kids peer in for a closer look, whereas most back off after they see the crayfish's claws flailing around, ready to pinch anything that gets close. "Only five kids have gotten pinched today, so you don't have to worry," I say as a joke. Even more students back away. Every student is laser-focused on the crayfish's claws as it flings them back and forth searching for something to pinch. I quickly put my finger in the danger zone, only to pull it out right before it gets pinched. With over fifteen years of cray-handling experience, I can usually predict when the crayfish will pinch and avoid any blood being drawn.

Crayfish claws are like a pair of scissors; when the two parts come together, a crushing force is generated thanks to the meaty muscle hidden under the claws' exoskeleton. One student that has been glued to the crayfish and whose nose is practically touching the crayfish's antennae asks, "How bad does it hurt to get pinched?"

"Would you like to find out?" I ask with a knowing smile. Without saying a word, the kid's head shakes convincingly side to side in small movements, which shows his childish fear. "Well, if you won't, I will," I say, as I inch my finger back into the danger zone.

Most crayfish pinches aren't that bad, but it depends on the size of the crayfish. Bigger crayfish with bigger claws produce more painful pinches. With only a few minutes left before the students move on, I slip my pointer finger into the sweet spot of the crayfish's claws. The second it senses my finger (thanks to the sensory hairs on the claw), the crayfish clamps down on my flesh. A few students look back at their parents, who are just as wide-eyed as their children.

Unfortunately for me (but fortunately for the students), the crayfish I was working with was an adult male—capable of pinching strongly enough to force my face into a grimace. Like before, half of the students are amazed and peer in for a closer look, while the other half back off and observe from a distance. While in the grip of this crayfish, I discuss the importance of crayfish claws. "They are like the hands of the crayfish, which are used to obtain food and scare off predators and sometimes are used as weapons during a fight," I say.

Fighting with other crayfish doesn't result in losing tablet privileges like it would if one of the children surrounding me fought with a sibling. For crayfish, fighting is a normal, routine activity. Having large, strong claws is a trait that may increase the crayfish's chance of winning a fight.

As the students leave, the crayfish's claw muscles tire from a long day of finger-pinching. I place my hand in the nearby aquarium and the crayfish immediately loosens its grip and tail-flips away. A quick look at my pointer finger reveals my battle scars. There are at least half a dozen pinch marks from a full day of painful education. Group after group, year after year, I sacrifice my finger at this outreach event in the name of science. Repeat visitors get the same spiel, but the students always remember my lesson about claws and the aggressive behavior of crayfish. The trade-off in education for personal pain is one that I would take any day.

ACROSS THE MORE THAN 700 species of crayfish, the spectrum of sizes, shapes, and colors of their claws is like a seasoned mechanic's toolbox: different tools for different jobs. Some claws, like those of the Shrimp Crayfish (*Faxonius lancifer*), couldn't intimidate a toddler, as they are puny, thin, and fragile looking. But some species like the Longpincered Crayfish (*Faxonius longidigitus*) have claws so long that they look like they are permanently carrying two unwieldy pairs of scissors. Other crayfish claws are just plain weird, like those of the New River Crayfish (*Cambarus chasmodactylus*), which have a teardrop-shaped hole inside smooth claws (I have a tattoo of this species).

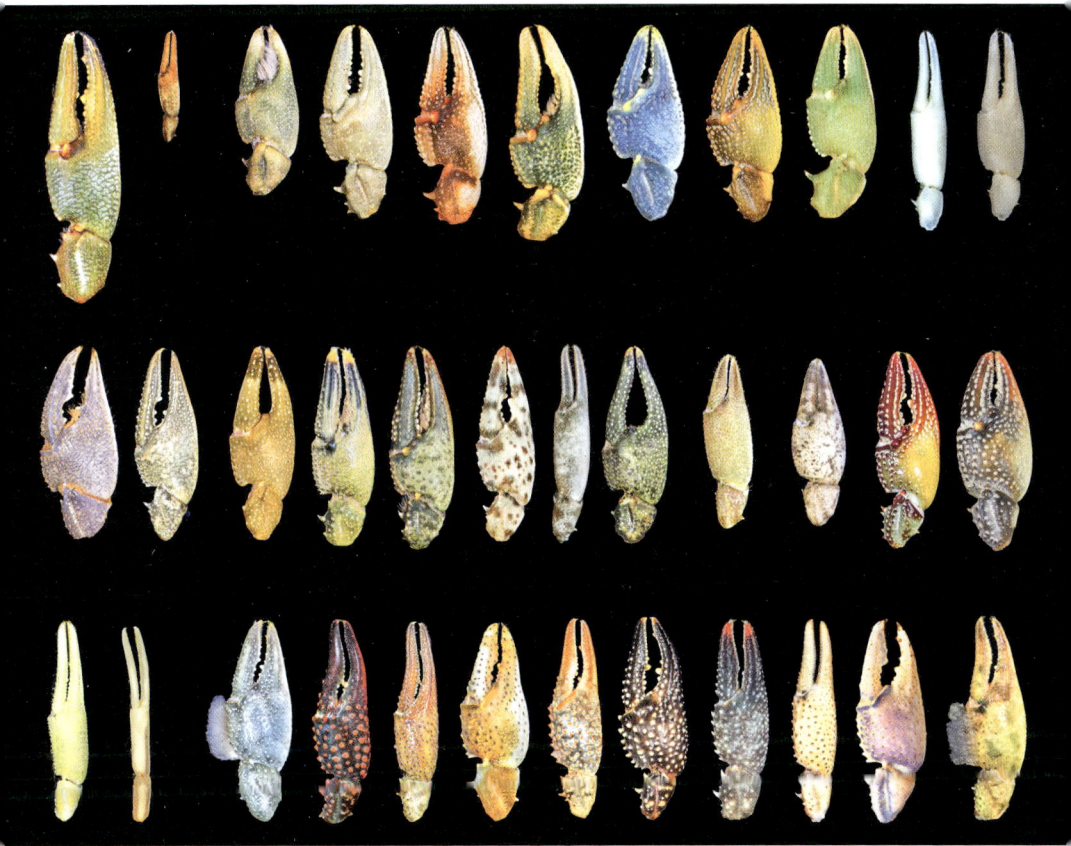

Variation in the size, shape, and color of crayfish claws across different species.
Photo by Guenter Schuster.

At their extreme, the claws of some crayfish species can be the same length as the rest of their body and weigh up to 40% of their entire body mass. (For me, this would equate to each of my arms weighing about forty pounds.) Claws, which are used for pinching, waving, and gallivanting about, are akin to the antlers of an elk or the horns of a bighorn sheep; they are first and foremost weapons used in fights. Whether the crayfish is standing its ground underwater or being used at a scientific show and tell, claws are used to deter predators or fight with other crayfish. Fighting is just as much a part of crayfish life as finding food, running from predators, and reproducing.

The aggressive tendencies of crayfish are evident in their behavior in the laboratory, which mimics what is observed in nature. Aquarists who have tried to keep multiple crayfish in a single aquarium without ample hiding spots know what I mean. Two crayfish go in, but only one comes out. Crayfish are unrelentingly territorial.

Defending a territory is not the only thing a crayfish needs to do to succeed, though; it needs food to persist, shelter to protect itself, and a mate to pass on its genes. But none of these resources are laid out buffet-style; they are scarce and the competition is fierce. In a given stretch of stream, there may be only one perfect-size rock under which to take shelter but a half dozen crayfish who want it, and they'll all fight to claim the shelter's protection. Contrary to the standard animal stereotype that males are the brutes, crayfish of both sexes are ready to brawl at any moment. Crayfish do not discriminate in whom they are willing to brawl with. Males fight males, females fight females, and males fight females. Crayfish of all age classes also fight, whether it's the oldest or largest crayfish in the stream or a juvenile who has been separated from their mom for only a few hours.

What amazes me the most about crayfish competition is how most interactions do not get resolved by powerful pinches. Rather, crayfish will first communicate with each other and engage in an intricate ensemble of escalated events to try and resolve the contest before they resort to barbaric combat. As such, in only extreme cases do crayfish injure each other; they try their best to deescalate a fight well before injuries occur. There are fringe cases, though, but serious injuries are rare.

Aggression in crayfish is typically more akin to a ballroom dance, with dozens of steps that follow each other in intricate fashion, before resolving gracefully. Despite what dramatized documentaries depict, aggressive interactions in nature rarely reach dangerous, injurious combat. Crayfish and other animals are smart. Instead of going straight to throwing punches or pinching off limbs, they try to talk it out. And in the case of crayfish, talking it out means spraying each other's faces with urine.

THROUGHOUT MY PHD, watching the escalation of events occurring during crayfish combat was an everyday endeavor. Sleepless nights and early mornings burdened me for years as a student, but watching crayfish fight was like a reality show and pro wrestling all in one, filled with a roller coaster of drama, rising action, a climax, and finally, resolution.

To tune in to the show, all you need is an aquarium and some crayfish. Since fighting is a part of their everyday life in the wild, they were more than willing to throw their claws around in the laboratory right in front of me—a fact I found delightful as a data-hungry doctoral student interested in understanding the role that communication plays in aggressive interactions.

Two crayfish in a territorial dispute. Photo by Chris Lukhaup.

Crayfish are model organisms for understanding how animals communicate during aggressive interactions, because they use a variety of visual and chemical signals to resolve territorial disputes that can be easily observed in the lab. This fact, along with my childhood interest in crayfish, led me to study these animals for my dissertation research.

Everything starts with an arena: an aquarium with a divider that can split the area in two. An ideal divider is clear but has small holes throughout, which provides a physical separation but not a chemical or visual one. The divider allows two crayfish to be placed on separate sides of the aquarium, where they can spend time acclimating to their surroundings and not be thrown straight into the heat of battle.

By imagining the crayfish umwelt, a better understanding can be gained of what goes on in these potentially injurious interactions. After being placed on one side of the arena, with a moving blur (the opponent) in the distance, the crayfish first explores, tapping and waving its antennae across every pebble and wall. All of a sudden, the antennules start to flick rapidly like a kid playing the drums. The scent of another crayfish becomes detectable. During this stage, each crayfish can learn a lot about the individual on the other side of the aquarium. Through smell alone, crayfish can discern whether their adversary is male or female and even whether it has previously engaged in a fight with them.

The games begin once the divider is lifted, but don't expect anything crazy right off the bat. Although crayfish are adept fighters, it pays off to ease into things and to make sure opponents don't pinch off more than they can chew. With the entire arena now free to explore, within a few seconds the opponents will meet for a formal greeting. Once physical contact is made (typically antennae-to-antennae), the combatants assume an aggressive posture: legs raised as if they are on stilts, with claws spread open to try and intimidate one another. Now in a staring contest, their antennae form a V and they take turns tapping each other with their antennae—trying to gauge the size, strength, and fighting ability of their opponent. Gentle antennal taps are contrasted with the rapid movements of the antennules, still trying to sniff and gain as much information as possible. Meanwhile, their claws become interlocked as if they are teasing each other with the statement "One wrong move and I will pinch you." Early pinching is poor manners—like a sucker punch (or rather, a sucker pinch), which is rare at this stage of the fight.

This "feeling out" stage, rife with interlocked claws, antennal taps, and close-distance antennule sniffing, can last anywhere from a few seconds to half an hour. Once a crayfish feels that the resource (in this case, an empty arena) is not worth the risk of fighting over and potentially getting injured, it waves the white flag, which involves making multiple rapid antennal taps over the top of the victor and lowering its posture, indicating that it is forfeiting this encounter. A forfeit can occur in any fight but is most common in fights where there is a large size discrepancy between opponents. The bigger the discrepancy, the shorter the fight.

The real show begins when the opponents are evenly size-matched and equal in their fighting ability. When this occurs, their interlocked claws go through several rounds of pushes, shoves, and strikes, like a ritualized back-and-forth arm wrestle. This phase is the equivalent of a pushing match between two combatants in a bar fight. Pushes (in humans) or claw strikes (in crayfish) cannot cause serious harm, but they are a good way to feel out how strong your opponent is. Throughout this claw activity, both crayfish also start to pour out bursts of urine. If the competitors can't figure out who is stronger through the pushing match, sending out urine and smelling the opponent's urine can help add that additional bit of information. Most, but not all, fights get resolved during this stage.

But as if a switch goes off in each crayfish's brain, the claw strikes and urine release can quickly transition into a full-out brawl. This switch is flipped in only the rarest occasions. But once flipped, the crayfish beat their antennules so fast that they become a blur, and with both crayfish pinching down on their opponent's claws, they go flying like professional acrobats as they try to coerce their opponent into giving up. Sporadically, one crayfish will lift its claws up and toss them backward like a forklift. Crayfish can also use "offensive tail flips" during this stage. Offensive tail flips are not designed for escape but rather are used to maneuver the crayfish's opponent into a better position so the crayfish can twist and turn its opponent to eventually gain an advantage. And in these moments, the urine floodgates open. Some crayfish give up immediately once the fight escalates to this stage, but others go on for hours (although small intermissions occur every few minutes; this likely gives the crayfish time to process the chemical information released by their opponent).

The winner ultimately is determined based on each crayfish's assessment of its fighting ability in relation to its opponent. Whichever crayfish waves the white antennal flags first is the loser, and to the winner go the spoils. The winner obtains whatever resources are being fought over. In nature, shelter, food, and mating opportunities are determined through this multistep, ritualized dance of aggressive behaviors. Each step in the fight has its role, and as the crayfish transition from one step to another, they glean more information about their opponent. A decision must eventually be made as to who concedes and who wins.

If the crayfish recognizes the urine of a crayfish it has beaten in the past, it will lunge at its opponent to reinforce its position of power. If it recognizes the urine of a crayfish it has lost to, it will disengage from any signs of aggression from the more dominant crayfish. Crayfishes' chemical memories have been shown to last anywhere from a few hours to several days.

A crayfish's ability to pee, interlock claws, and fight is tied to its success. Disputes over limited food and shelter are paramount for a crayfish to avoid starvation or hide from predators. But for crayfishes to have become one of the most successful freshwater organisms in North America, they needed to do much more than survive; they needed to reproduce. And for crayfish, sex and fighting go hand in hand.

## Sex, Love, and Fighting

Once I was convinced that crayfish were the ideal system for my research, I would drop a few crayfish into a container and observe their behavior, trying to get into the world of a crayfish. My "laboratory" at this time looked like the back shelves of a thrift store: wire racks covered in Tupperware containers, some old aquariums, and even older desks saved from the trash.

In one specific instance, I was practicing watching male crayfish fight to ensure that I could predict and identify the series of events that occurred in real time. In just a few months, I would have to start collecting data from hundreds of interactions—I wanted to make sure nothing I saw would be a surprise. My typical setup was right in front of me: a glass aquarium, some gravel, and a crayfish on each side of a divider. After lifting the barrier and watching the crayfish interact, I saw the same typical series of events that occurred in any crayfish fight: antenna taps, antennule flicks, and claw interlocking. Everything as expected.

But an unexpected email notification took my attention away for a few minutes. After taking thirty seconds to respond to the email, I looked up to see a position I had not seen in other fights: The two crayfish were in a tight embrace, stomach to stomach—missionary position.

Before setting up these encounters, I would normally peek at the underside of the crayfish before dropping them into the aquarium. Crayfishes from North America all have identifiable organs on their underside that can be used to determine their sex. Males have a pair of intromittent organs called gonopods (Latin for "sex feet") that function as dual penises at the base of their abdomen, whereas females lack these organs and instead have a sperm storage receptacle, called an *annulus ventralis* (Latin for "belly ring"). Male North American crayfishes also have a pair of hooks on their walking legs to pin the female down and ensure that their sexual parts properly connect, whereas females have openings called gonopores, from which eggs are released.

I had assumed that the crayfish in my aquarium were both males, but a mislabeled Tupperware container led me to set up a crayfish date right in front of my eyes. This was the exact type of mistake I was trying to learn from before I had to do serious data collection. But I was far from upset, as it was the first time I had witnessed this embrace, which I had read about a dozen times. Consistent with what I had read, locked in the act, the male was on top and the female on the bottom—which is standard in all crayfishes except for Asian species in the genus *Cambaroides*, where the female is on top. And while the male is on top, a few thrusts ensure that the male gonopods are firmly inserted into the female's annulus ventralis.

Before a female crayfish will accept a male crayfish as one of her mates, they go through a pre-fight sex ritual. This ritual is what I observed in my lab and mistook for a dispute between males. Female crayfish want to mate with a strong and capable male crayfish, so they fight with them to test their skills. And once the female accepts the male's fighting abilities and deems him a suitable mate, her body turns stiff; this triggers the male to flip her into position. Poor fighters or small males are rejected by females and end up mateless.

A mating pair of crayfish. Photo by Chris Lukhaup.

BIOLOGISTS KNOW A LOT about crayfish sex. Without a strong under-standing of reproduction and mating in crayfish, it would be difficult to manipulate the crayfish breeding cycle to maximum production for human consumption (see chapter 6).

Aside from humans and crayfish, few animals have sex in the missionary position: Some primates, some cats, and some cetaceans (dolphins and whales) do as well, but it is a rarity in the animal kingdom. (Presumably because of how exposed both participants are in this embrace.) The crayfish missionary embrace generally lasts a few minutes, but some species have been reported to mate for hours! Regardless of time, the males engage in several rounds of thrusting to ensure that sperm is transferred. After the act is completed, both crayfish may go on their way without ever interact-ing again, and the female is left to care for the young crayfishes that will hatch from eggs in anywhere from one to a few months. Some males even leave a present for any future males that try to mate with that female: They leave a sperm plug that blocks the female's sperm-storage organ from being filled up a second time by another male. And if things weren't already weird enough, another male can come along and eat this sperm plug as a snack before depositing his sperm. But sometimes the plug works as a deterrent for a lazy male.

In the wild, crayfish are promiscuous. Both males and females mate multiple times. The largest males are capable of mating with the most females, as they are seldom rejected, unlike a small male with smaller claws. Females also prefer males of a large body and claw size, as opposed to a puny crayfish that is missing his claws—showing his battle scars from many lost fights.

While the males are busy fighting and looking for more females to mate with, mated females search for a safe location to extrude their eggs and begin motherhood. Before the female releases her eggs, she goes through a hormonal change in her body that leads to the development of white glands on the underside of her abdomen. These glands, called glair glands, are often referred to as glue glands, as they secrete a sticky, glue-like substance to hold the eggs together after they are extruded. Without the glair, the eggs would likely drift off into the water and be unable to get the care they require.

Even after copulation has occurred, fertilization of egg and sperm has not, because the female has received the sperm only in her annulus ventralis. And one female might have half a dozen packets of sperm, called spermatophores, from half a dozen lucky males. When a vertebrate receives sperm, fertilization occurs internally—all the magic happens automatically, and it

The underside of a female crayfish exhibiting active glair glands. Photo by Zackary A. Graham.

An ovigerous "in berry" female crayfish. Photo by Zackary A. Graham.

is just a waiting game until the young arrive. But in crayfish, once the female receives spermatophores, she is tasked with following a few steps, as if she is reading a recipe, to ensure that fertilization occurs.

Fertilization occurs when the sperm comes in contact with the female's eggs; the sperm and eggs are both stored internally, in separate, unconnected pouches. It is the female's job to make sure that these sex cells are extruded and intermingle. When the time is right, the female flips onto her back and uses her stomach and tail as a workbench. One by one, her eggs are extruded from her gonopores underneath her third walking leg. During egg extrusion, the glair glands secrete the gluey glair. The glair softens the sperm inside the female's annulus ventralis, which is then also extruded. Now, the sperm and eggs all mix in a kind of fertilization soup, thanks to the glue-like glair, and just like that, the eggs are fertilized. One by one, eggs pop out of her legs and mix into the fertilization soup, and then the female crayfish's eight pairs of walking legs shift the eggs beneath her abdomen, where the eggs attach to structures called pleopods for safekeeping.

Depending on the species, crayfish can lay anywhere from fifteen to over 1,000 eggs in a single event. The eggs are packed tightly together in a massive cluster. Females carrying eggs at this point are often called ovigerous ("egg carrying") or "in berry," because their jumbled cluster of eggs resembles a massive blackberry under their abdomen.

After fertilization has occurred and the eggs are tucked underneath the abdomen for safekeeping, a female crayfish searches for a safe and permanent shelter where she can get some alone time to care for her young—away from annoying males or dangerous predators. Soon, she will be overcome by potentially hundreds of baby craylings. During this time, she will also use the movable pleopods to fan her eggs to increase aeration and ensure the eggs do not run out of oxygen. During this entire process, female crayfish will seldom eat—only surviving off of stored energy consumed in the months leading up to motherhood. Wandering out with potentially hundreds of eggs tucked underneath your abdomen is far too big of a risk.

After a few weeks of being protected underneath mom's abdomen, the craylings will progress through their development in hopes of hatching. If you find an ovigerous female at the right time of year, you can even see two tiny dots on each of the hundreds of eggs, which are the eyes of the developing craylings.

Soon, each crayfish will emerge from its egg; at this point, each of the crayfish looks like a tiny adult crayfish the size of a Tic Tac. Lucky for mom, she has no risk of losing track of her babies at this point, because each of them is leashed to her with a telson thread, an ingenious adaptation to life in flowing water. It's a thin umbilical-cord-like structure that attaches each crayling's tail fan to the egg case, which is still attached to the mom.

In the early stages after hatching, craylings are completely reliant on their mother for protection. She keeps them safe in her chosen shelter and allows them to consume particulate matter off of her and in their immediate surroundings. After molting several times, the craylings will eventually grow and become confident enough to venture out into their immediate area with their mother, crawling around her like a swarm of underwater flies.

The craylings go through four major molts, through which they become stronger and more capable of living without their mother. After the fourth molt, a crayling has everything it needs to survive on its own: claws for grabbing and pinching and an abdomen for tail-flipping away from predators. The craylings sometimes want to stick around and continue to mooch off their mom for protection, but if they stick around too long, they might end up becoming dinner. Early on in development, the craylings release a chemical that greatly reduces the mother's willingness to cannibalize her new sons and daughters. After all, she just spent the past few weeks or months developing and caring for her young, all without ever venturing out to eat a good meal on her own. But after the fourth molt the craylings no longer produce that chemical, putting them at risk of being cannibalized.

Risk of cannibalization aside, crayfish are exceptional mothers, especially for invertebrates, which are notorious for producing thousands of offspring and then never looking at them again. Female crayfish dedicate a few months a year entirely to raising their offspring, tolerating and protecting dozens of little versions of themselves that use their bodies like a jungle gym. In fact, the caring nature of maternal crayfishes is thought to be a key trait that has led to their freshwater success. Their marine lobster cousins, on the other hand, may lay 100,000 eggs at once, but once the eggs hatch the mother kicks the baby lobsters out immediately and provides no care after hatching. Contrast this with crayfishes, which let their young stick around to molt four times before they leave the nest.

WHEN IT COMES TO REPRODUCTION, not all crayfish are created equal. Smaller species, like the festively named Christmas Tree Crayfish (*Procambarus pygmaeus*), are known to produce only two or three dozen eggs at a time, whereas others, like the aptly named Virile Crayfish (*Faxonius virilis*), carry more than 500 eggs underneath their abdomen. The speed and number of reproductive events also vary. Over the approximately five-year lifespan of a Virile Crayfish, a single mother may extrude, develop, and care for well over 2,000 young. And if half of those young are females, it will only be a year or two for each of them to start popping out their own crops of eggs.

The ability of some crayfishes to grow quickly, reproduce often, and produce tons of babies is the trait that makes them important for natural ecosystems. But issues arise when these "super-reproducers" are moved outside of their natural range. Then they become competitors to other crayfish species that grow and reproduce slowly, which comes with disastrous consequences.

# CHAPTER 6

## Aquatic Invaders

### The Dark Side of Crayfish Biology

A NATIONAL PARK BIOLOGIST glares down at a haphazard cluster of large, clunky, squarish rocks on a lakeshore—peeling them back one by one to witness a flourishing population of Signal Crayfish (*Pacifastacus leniusculus*) darting away as they are revealed. This species has a hefty body, is rusty brown all over, and is well endowed in the claw department. They thrive in these rock clusters, which are the remnants of a volcano that erupted a mere 7,700 years ago. Only a hundred feet offshore from these rocks lies open water deeper than in any other lake in the United States, where rainbow trout and kokanee salmon search for wandering crayfish.

In the middle of the lake sits Wizard Island, the most prominent (and only) feature in the circular body of water. Zoom out and you will see that the entire lake is located within the confines of a natural bowl-like structure called a caldera. Its sheer, steep edges tower above the water, creating a natural barrier that makes escape impossible for the lake's aquatic inhabitants. In the almost perfect circle, the jagged incline is unbroken, leaving no avenue for water to flow out of the circular basin. The aquatic animals that inhabit this area are confined, with no (natural) way in or out.

The scene before you can only be found in one place in the United States: the remote and secluded Crater Lake, in Crater Lake National Park in southwestern Oregon. This stunning lake is nestled within a cauldron-shaped depression, formed by the volcanic eruption and collapse of Mount Mazama. The park rangers work hard to keep this national treasure in pristine condition. But the thriving population of Signal Crayfish beneath the rocky shores is a source of concern for the park's wildlife biologists. These crayfish, though flourishing in the environment, are not natural inhabitants of the lake. They are foreigners. They are invasive.

Humans brought Signal Crayfish to Crater Lake intentionally to serve as food for the fish, which were also introduced but lacked a food source. At first, Signal Crayfish numbers were minimal, and they could be found only in a few sections of the lakeshore. But year after year, Signal Crayfish spread, with some even being collected at depths of 250 meters (820 ft). And in these areas, they outcompete the native wildlife, including the Mazama Newt (*Taricha granulosa mazamae*), a rare amphibian found only in Crater Lake. It's a sobering reminder of the impact humans can have on even the most pristine environments.

Throughout this book I have highlighted the diversity and significance of crayfish. Until now, I have not confronted the dark side of crayfish biology. "Hate" and "destruction" are words I have not used in previous chapters. But all of that changes now. In their native ranges, crayfish have positive impacts on their environment. In areas that have evolved to maintain and support

Crater Lake, with a view of Wizard Island. Wikimedia Commons.

crayfish, they are a part of the natural balancing act of an ecosystem. But when moved outside of their natural range, things go awry.

Spreading into foreign habitats is far from a crayfish-specific issue, with many species, such as feral swine and Burmese pythons, regularly making headlines for the ecological destruction they cause when transported out of their natural range. And none of this is by their own means but rather because of humans transporting them around the globe. Not good. When crayfish get moved to a foreign habitat, they can devastate an ecosystem. It's becoming a worldwide problem.

# AN IMPORTANT DISTINCTION: NATIVE VERSUS NONNATIVE VERSUS INVASIVE

Every animal has a natural range where it belongs, the place where it originated, or what is called its native range. An animal's native range is bounded by barriers, like a mountain range or the extreme heat of a desert. Although a species may move around widely throughout its native range, it has its limits. Few species are natural globe-trotters.

In some scenarios, animals are moved outside their natural limits—often by humans. Anytime an animal is moved outside of its native range, it is considered a nonnative species (also referred to as an introduced species, nonindigenous species, or alien species). Anytime a nonnative species starts to spread outside of its initial nonnative range or starts to negatively impact the environment where it has spread, it is considered an invasive species. Invasive species are one of the top causes of biodiversity declines worldwide. Some of the most common species you are used to seeing outside (both plants and animals) may be considered nonnative or invasive, depending on where you live. With these definitions, biologists can make the distinction that all invasive species are considered nonnative, but not all nonnative species are considered invasive.

# The Ways Crayfish Spread

The natural dispersal of crayfish is limited. They tend to stay put in the body of water where they are born, with some exceptions, like overland migrations during rainy or humid nights. But add in some ill-advised *Homo sapiens*, and crayfishes' dispersal abilities become endless. Crayfish have been moved around by humans for hundreds, if not thousands, of years. Nowadays, you can flip a rock in the Great Lakes or the Hawaiian Islands and find the exact same species, the Red Swamp Crayfish (*Procambarus clarkii*), which is a long way from its home in the Gulf Coastal Plain from Florida to Mexico.

Why do crayfish get spread around? And who is doing the spreading? There is no single answer to these questions, but there are a few culprits that have been pinpointed to be the major spreaders of crayfish. Our journey through the dark side of crayfish biology begins with an examination of the curious and often alarming ways crayfish are whisked around the globe, intentionally or otherwise. From there, the narrative shifts to their disruptive impacts on ecosystems and the dilemma of whether these crustaceans can be stopped.

## FISHING AND BAIT-BUCKET INTRODUCTIONS

Crayfish, or crawdads, as anglers call them, are prized fishing bait. Depending on your area, the shelves of your local sporting goods stores will likely have a few crayfish lures that come in all shapes, sizes, and colors. Depending on your region, a well-stocked bait shop may have not only a selection of artificial lures but live crayfish for sale. But here's the catch: When live crayfish are used as fishing bait, anglers frequently transport them significant distances from the point of purchase. They may visit their local tackle shop for some crawdads before embarking on an hour-long journey to their favorite fishing hole dozens of miles away, transporting the crayfish beyond their natural range.

Many things can go awry once crayfish are transported. They can squirm off a hook or crawl out of bait buckets (I have seen it happen). Worst of all, if a fisherman fails to catch any fish or if he purchased too much bait, he may release the remaining crayfish into the water before heading home. It's always a good idea to use bait that's native to the area where you're fishing. It's never a good idea to use bait from outside areas. It's also important to keep in mind that some waters may have invasive species lurking in the depths—some of which you can accidentally transport from one place to

# THE CRAYFISH
# MOST WANTED LIST

Worldwide, there are somewhere in the range of twenty different crayfish species that have established invasive populations. In the United States, only a few species have been repeatedly spread around. Descriptions of the top invaders are below.

- **Rusty Crayfish** (*Faxonius rusticus*): This species is native to the Ohio River Basin in the states of Ohio, Michigan, Indiana, Kentucky, and Tennessee. This species now is invasive in more than twenty states, mostly across the Eastern United States, but a few invasive populations exist to the west. This species is easily identified by the characteristic purple-brown patch on the sides of its carapace.

- **Virile Crayfish** (*Faxonius virilis*): The Virile Crayfish has a wide-ranging native distribution, including the Great Lakes and Upper Mississippi River Basin, occurring natively in most of the upper regions of the Midwest. Invasive Virile Crayfish are now widespread throughout the United States and can be found in the majority of states. This species is highly variable

A Rusty Crayfish (*Faxonius rusticus*). Photo by Guenter Schuster.

A Virile Crayfish (*Faxonius virilis*). Wikimedia Commons.

across its distribution, but they often wield blue-green claws with large yellowish tubercles (hard bumps) that are distinct from their brown bodies.

- **Red Swamp Crayfish** (*Procambarus clarkii*): Red Swamp Crayfish are native to the Gulf Coast states but creep up into the lower reaches of the Mississippi River Basin. This species has popped up in invasive populations around the country and is difficult to detect and remove because of the crayfishes' ability to burrow. They are one of a few bright-red species but are also characterized by numerous spines and bumps all over their body.

- **Signal Crayfish** (*Pacifastacus leniusculus*): Native populations of this species are found in the Columbia River Basin in the Pacific Northwest and into Canada. This species is adapted to colder waters, which has contributed to its success as an invasive species in many areas of the northwestern portions of the United States. They are especially common invaders in California. Signal Crayfish are typically brownish on top with bright-red markings underneath their claws. Most populations also have a white "signal" spot on their claws.

I encourage everyone to become familiar with these species and other potential invaders in their area. Depending on where you reside, these species may be the most likely crayfishes you will encounter. A Google search for "invasive crayfish in [your location]" should provide educational material specific to your area.

A Signal Crayfish (*Pacifastacus leniusculus*). Photo by Guenter Schuster.

A Red Swamp Crayfish (*Procambarus clarkii*). Wikimedia Commons.

another. So before you cast that line, make sure you know what you're dealing with. One careless act of dumping leftover crayfish from your bait bucket into a new environment can have permanent consequences.

In North America specifically, the dispersal of nonnative crayfish through fishing and bait-bucket introductions is likely the number one cause of the spread of invasive crayfish. Areas with high densities of fishermen, like the Great Lakes or the hundreds of isolated lakes of northern Wisconsin, are hotbeds for invasive crayfish introductions. Species in the genus *Faxonius*, like the now widespread Rusty Crayfish (*F. rusticus*) and Virile Crayfish (*F. virilis*), are top prospects for fishing bait because they grow quickly and reproduce easily in captivity. Both species have similarly been introduced into dozens of locations throughout the United States, especially in the Midwest and mid-Atlantic regions.

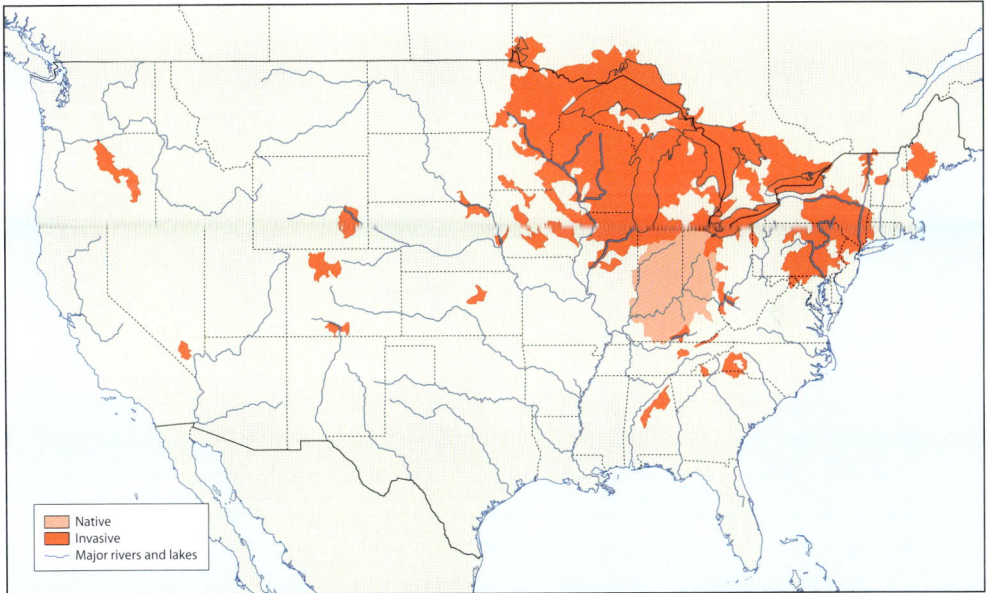

Native and invasive range of Rusty Crayfish (*Faxonius rusticus*). This map depicts the native range (light orange) and invasive range (orange) of the Rusty Crayfish, a species that has been spread across the United States. Isolated introductions have occurred all over the States, with this map depicting a small minority of known introductions.

## AQUACULTURE AND CRAYFISH FARMING

To grasp the enormity of the next crayfish-spreading vector, one must delve into the past. French colonists occupying Nova Scotia and New Brunswick, Canada, were expelled in the late seventeenth century by the English and settled down in what is now known as the state of Louisiana, where they took root and blossomed into the Cajun culture. Now, the term "Cajun" is used to refer to the entire culture of this region (mostly southern Louisiana) and less so to the initial French colonizers. But upon the colonists' arrival, this region saw an immediate influx of culture: from architecture to language, music, and of course, cuisine. Perhaps no other foodstuff captures the essence of the Cajun spirit more than the humble crayfish (regionally known as crawfish). The marine lobsters that the Cajuns were used to up north had been abundant and delectable. But when they moved south, lobsters were nowhere to be found in the bayous of Louisiana. So the Cajuns adapted and turned their culinary affections to lobsters' smaller, but more plentiful, freshwater cousins.

It should be noted that Indigenous peoples had been gathering crawfish in this region long before the Cajuns arrived, but it was the latter who cemented these crustaceans as a regional delicacy. Spend a few days in Louisiana, and you'll soon realize that crawfish are woven into the very fabric of the state's identity. Head to any grocery store and you'll see glistening packages of peeled crawfish displayed front and center in the chillers. And come summer, you'll find posters advertising a plethora of crawfish festivals, where folks from all walks of life gather to partake in the communal consumption of crawfish.

Whenever I mention that I study crayfish, there's one question I always expect: "So . . . do you eat them?" It's understandable, really. Crayfish have a strong culinary reputation. But despite dedicating a significant portion of my life to understanding these creatures, I've never actually tasted one. Not because I'm squeamish or harbor any moral objections but because I'm, rather ironically, deathly allergic to them—and to shellfish in general. This rather inconvenient condition reared its head in high school, when I learned about it the hard way after a few unfortunate encounters with cocktail shrimp. Eating them led to some throat-swelling excitement I'd prefer not to repeat. Handling crayfish doesn't bother me in the slightest, so I can work with them all day without a problem—as long as none of them end up on my dinner plate.

I'm more than happy to champion crayfish as a delicacy for others, though, appreciating the cultural and economic roles they play. On camping

Crawfish festivals attract many people in the Southern states, especially in Louisiana. Wikimedia Commons.

A crawfish farm in Louisiana. Photo by Joseph Connors IV.

trips, I'll even boil a few in a tin pot over the fire, giving my friends a running commentary on crayfish anatomy as they prepare to dig in. But I'll always be an enthusiastic observer, not a participant, in the feasting.

Now, when it comes to crawfish consumption around the world, there's one species that reigns supreme, and it's the Red Swamp Crayfish. If you have ever consumed a crawfish, it was likely this species. They're fast growing, fecund, and thrive in the coastal bayous and swamps that run from Mexico to Florida, all across the Mississippi and Gulf of Mexico drainages. In Louisiana, where crawfish are practically a religion, the vast majority of the wild harvest of this species is pulled from the Atchafalaya Basin. Here, the natural flood pulses of the basin are perfectly timed with the Red Swamp Crayfish's reproductive cycle.

The wild crawfish harvest remained local to Louisianians for quite some time until the commercial sale of crawfish picked up in the 1800s. But the lack of transportation and ability to keep crawfish meat cold kept these animals a local delicacy. Once cold trucks and transportation on ice were

# CHINESE DELICACY

For most of the history of crawfish farming, the United States, and specifically Louisiana, was a top producer. But now things have changed and 90% of the crawfish in the world are consumed in China. What's more, the time of the United States dominating world crawfish production is long gone, as China is now responsible for an estimated 70% of it, primarily by using the dual-crop system of farming crayfish and rice. The Chinese government even provides subsidies to small-pond crawfish farmers to promote the sustainability and growth of this expanding market.

The rapid integration of crawfish farming and consumption into Chinese culture has taken these animals to far-flung reaches of the world that they have previously never been in. Crayfish are so popular that even American fast-food giants in China sell crawfish-inspired cuisine. You can grab crawfish tacos at Taco Bell or crawfish pizza at Pizza Hut, and you can even find crawfish-flavored Lay's potato chips. The most popular serving method is with a "thirteen spices" mix, which is a fiery oil-and-seasoning-based mixture.

figured out, crawfish could be shipped all around, which timed up perfectly with advent of aquatic farming of crawfish, which forever changed the game for the Red Swamp Crayfish and other popular commercial species.

Wild crawfish harvesting from the Atchafalaya became less and less lucrative due to fluctuating prices and the development of a more cost-effective method of catching crawfish. Now, in Louisiana, which is where 90% of crawfish are produced in the United States, a mere 10% of the annual crawfish harvest comes from the wild. The remainder comes from large aquaculture ponds where Red Swamp Crayfish (and occasionally other species) are introduced, left to do their business, and then collected with baited traps.

The productivity and ease of crawfish farming made it more cost effective than catching wild crawfish. Crawfish farming exploded even further when farmers realized they could use the same pond to alternate between growing crawfish in one season and rice in the next. It turns out that this dual-crop system not only raises profit margins but can also stabilize potential market shifts in the price of both goods. It's a win-win situation.

A farmer takes a waterlogged patch of land and sows it with rice. As the rice grows tall enough to provide a bit of shade and shelter, the farmer introduces crawfish into the mix. The crawfish consume some of the rice crop, but there is still enough rice at this point for a decent harvest. In turn, the crawfish waste fertilizes the soil for the next round of rice, making this a symbiotic system of farming. Similar systems of rice/animal farming have been implemented before, but none were quite as successful the crawfish and rice combo, which soon became a worldwide affair. The potential economic and food security boost that comes from growing both a carbohydrate-rich plant and a protein-rich crustacean in the same field took hold and is one of the principal reasons that crayfish have been spread around the world.

This ingenious system of farming has become a global sensation, with the Red Swamp Crayfish leading the way. This fast-growing, highly reproductive species has become a staple of crawfish farming, which has spread like wildfire to forty countries across four continents. In the United States alone, thirty-eight states have introduced populations of Red Swamp Crayfish. It's hard to overstate just how successful and widespread this red crustacean has become. Of course, there are other species popular in the crawfish-farming world, such as the Red Claw Crayfish (*Cherax quadricarinatus*) and the Yabby (*Cherax destructor*). But the Red Swamp Crayfish reigns supreme.

Everything suggests that the Red Swamp Crayfish is the most numerous and widespread species in the world. Because of this, they are perhaps the most feared and prevalent of all crayfish invaders. Even when kept under close watch, this species is capable of burrowing deep down into soils and can evade detection. And this species is surprisingly adaptable, being able to survive in all but the most extreme of climates. In Louisiana, there have even been several reports of massive land migrations on warm, rainy spring nights, where thousands of individuals migrate from one area to the next. Spreading is not an issue for this species. All it takes is a few crayfish to escape from a holding tank or one on-land mission during a rainy evening to get into an area they are not intended to be. This story has unfolded dozens of times in the United States and even more times globally.

## THE CRAYFISH PET TRADE

Walk into any big-box pet store and not only will you see exotic fish shipped in from African lakes or Amazonian streams, but you may also see some curiously named crustaceans like an "Electric Blue Lobster," "Miniature Orange Lobster," or some other hodgepodge of two adjectives tacked onto "Lobster." This should be no surprise by now, but these animals are fresh-

A popular pet crayfish kept in aquariums, the Dwarf Mexican Crayfish (*Cambarellus patzcuarensis*) is named after Lake Pátzcuaro in Mexico. Although it's among the most common crayfish sold around the world, this species is threatened in the wild. Wikimedia Commons.

water crayfish erroneously labeled as lobsters. In the United States, the crayfish species you are most likely to see for sale are color variants of North American species that are typically less colorful but are selectively bred for their blue or orange hues or even koi-like patterns of orange and white. For example, "Electric Blue Lobsters" are a blue variation of the Everglades Crayfish (*Procambarus alleni*), whereas "Miniature Orange Lobsters" are an orange morph of the Dwarf Mexican Crayfish (*Cambarellus patzcuarensis*). In the wild, both species have muted brown and tan coloration.

Online, you'll find even more attractive names and multicolored species for sale, such as the Thunderbolt Crayfish (*Cherax pulcher*) or the Supernova Crayfish (*Cherax boesemani*), which are beautifully colored species in the genus *Cherax* from the crystal-clear waters of Papua New Guinea. The brighter and more conspicuous the coloration, the more marketable they are to potential enthusiasts.

In the United States, the market for ornamental crayfish and other crustaceans such as shrimps and crabs is relatively mild. If you want to purchase a pet crayfish in the United States, you are unlikely to find more than a few standard species for sale. Worldwide, the attraction to pet crayfish is much greater, which has led to some of the rarest US crayfishes being collected (often illegally) and shipped (also illegally) around the world for the pet trade. Every so often, a crayfish-smuggling attempt is foiled by customs officials, and an expert astacologist is called in to identify the contraband.

This trend picked up steam in the 1990s, when Europeans developed a keen interest in keeping crayfish as pets. In Germany specifically, where there are only three native crayfish species, the market exploded. A 2013 study revealed that 123 different species were found in the German ornamental crayfish trade, 104 of which came from the United States. And in the years since, an average of seven new species have been imported annually. The most sought-after species tend to be the most naturally colorful and uniquely patterned. But in many cases, hobbyists have been able to breed some naturally less-desirable crayfish into variations that appear ghostly white or have polka dots. I spend most of my lunch breaks perusing online crayfish forums, where people upload images of their collection of crayfish pets, and am always surprised by what selective breeding can produce.

As of recently, the world of ornamental crayfish has become popular in Asia, particularly in China and Taiwan. The appetite for collecting and breeding the most crayfish matches the appetite to consume them en masse. While googling rare crayfish species I know little about (something I also do on my lunch breaks), I often wander onto foreign websites and Facebook groups that plaster photos of the crayfish I am searching for and which to my surprise are being bred and sold by crayfish hobbyists. In many cases, the species showcased on these sites are threatened or endangered at the state level, which means they are being illegally collected and shipped around the world. In fact, some biologists are starting to conceal the collection locations of these species in the scientific literature to prevent the harvesting of natural populations for the pet trade. For other exotic animals like reptiles and marine fish, there is a long and perilous history documenting how the pet trade leads to the decline of native populations.

To date, only one non-American crayfish species has been introduced into the United States, and that is the Red Claw Crayfish. Red Claws are native to Northeastern Australia and the remote but crayfish-filled islands of Papua New Guinea. Because this species is 1) objectively beautiful and kept as pets and 2) a fast-growing species used in aquaculture, they have been moved around the world to warm tropical climates that match their native habitat. There are only three known populations of Red Claws in the United States. Isolated introductions occurred in Brownsville, Texas; Los Angeles, California; and Las Vegas, Nevada. This species has also been introduced throughout Puerto Rico.

Fortunately, in the world of the crayfish pet trade, the average hobbyist who partakes in this pursuit is often knowledgeable about the creatures they keep. Unlike your average buyer, these enthusiasts have taken the time to understand the specific requirements of each species. It is worth noting that while the pet trade has played a role in the spread of invasive crayfish, the number of introductions may be low compared to other vectors like aquaculture or fishing.

However, it's important to acknowledge that recent developments may alter this trend, as one particular species of crayfish has been identified as having the highest invasive potential of any crayfish (detailed below). This discovery could change the game and may render my earlier statements obsolete. Only time will tell what the future holds for the ornamental crayfish trade.

In theory, moving crayfish around from aquarium to aquarium is harmless. But all it takes is a few uneducated mistakes to lead to serious environmental issues. The typical threat starts when a child or parent decides that they no longer want to care for their pet crayfish. Their intuition leads

# KEEPING CRAYFISH

Although the crayfish pet trade has caused harm when introductions of invasive species occur, I believe that there is still a benefit to this industry. Keeping crayfish teaches you about their biology, and once you start learning about them, it's hard to stop. Some crayfish biologists even got started through their initial experiences with keeping pet crayfish.

I suggest that if you are keeping a pet crayfish for yourself or a child, that you catch a crayfish locally. Do some research first and find out what species are around you, and there will likely be a species amenable to being kept in an aquarium. A single crayfish can thrive for years in a five-to-ten-gallon aquarium with a filter. Having a lid is critical. I can confirm that if there is a way for a crayfish to escape, even the smallest of cracks or openings, it will. Alternating between dead leaves and standard invertebrate pellet food will keep the crayfish on a balanced diet. If you go down this route, remember that you should never put the crayfish back into a different body of water, and many suggest that you should not put it back in the environment at all because of the risk of disease spread.

them to believe that "a crayfish is a crayfish" and that these animals are all the same. So they feel obligated to release the crayfish back in to the wild "where it belongs." This can lead a crayfish from Texas to be thrown into New York's Central Park. These introductions come from sheer ignorance and a misunderstanding of the animals.

## MISCELLANEOUS VECTORS

Crayfish are primarily spread through fishing, aquaculture, and the pet trade. But a few oddities and outliers may also be vectors for the spread of invasive crayfish. These vectors are probably less common but still worth highlighting, because many come from unexpected places.

Golf courses, of all places, may be inadvertently contributing to the spread of invasive crayfish. The issue is that golf courses tend to create ponds as water hazards or to enhance their aesthetics. While these ponds may look beautiful when surrounded by the well-manicured greens, they can quickly become overrun by aquatic plants. To solve this plant problem, golf courses will introduce crayfish to the ponds to help manage the plant overpopulation. While this may seem like a harmless and well-intentioned solution, once the crayfish are inside the ponds, they may be able to escape and spread to other bodies of water.

In other situations, crayfish are also intentionally introduced into an area to solve a problem but later become an even bigger problem themselves. In the arid canals of Phoenix, Arizona, officials faced a conundrum when the growth of both native and invasive plants started to impede the flow of water. They needed a solution, and they thought they had found one in the form of the Virile Crayfish, native to the Great Lakes region. The crayfish were introduced to the canals with the hope that they would eat the plants and stop them from clogging up the canals. For a while, it seemed like a good idea. The crayfish did what they were supposed to do and chowed down on the plant mass. But then things took a turn for the worse. Surprise, surprise: The crayfish population started to explode.

It became clear that Phoenix needed to find a solution to the invasive crayfish problem, and officials decided to introduce a fish species that could eat the crayfish. The problem with this solution was that the species they chose, such as grass carp and tilapia, were also potential invaders. This approach may have served as a quick fix for the crayfish problem, but now all these invasive species were being spread even more. For example, the Virile Crayfish, which was originally introduced to canals in Phoenix, made it to northern Arizona, presumably from several bait-bucket-type introductions.

# AVIAN HITCHHIKERS

If I had to give an award for the most unique way in which crayfish may be spread around, it would be this one, which involves my favorite experiment ever conducted. This means of spread is a weird one, because it is technically natural (that is, without influence from "unnatural" humans), but nonetheless it is a rapid transportation method that can take a crayfish far from home—not thanks to humans but thanks to birds.

In a paper titled "Waterbird-Mediated Passive Dispersal Is a Viable Process for Crayfish (Procambarus clarkii)," a group of Portuguese researchers wanted to test the hypothesis that juvenile crayfish can use their tiny claws to latch on to a bird, like a duck or goose, and then hang on for dear life until they are dropped in a body of water outside their native range. In theory, this mode of movement could expand the native range of a crayfish species and propel them over impenetrable barriers like mountains or man-made dams. This experiment had two phases. In phase 1, which the researchers describe as "estimating the probability of crayfish clinging to a duck," they placed a fresh, dead domestic duck into an aquarium with ninety-eight juvenile crayfish. This initial trial was conducted to determine whether the juveniles would ever naturally explore enough to take shelter in the feathery matrix of the duck's feathers.

In phase 2, the researchers wanted to study crayfish under "simulated flight conditions." In this phase, they first tested whether crayfish could hold on to a mesh bag that they held out of the window of a car traveling 35–73 kilometers per hour (about 21–45 mph), which is comparable to the flight speed of pigeons and ducks. Then, they tied the mesh bag to the legs of trained homing pigeons to see if the crayfish could hang on while attached to a flying bird.

Despite their strange methods, their results are insightful and draw into question how crayfish are moving around. In the first phase of their experiment, they found that juvenile crayfish could cling to the feathers of birds. In the second phase, they found that the crayfish could hang on tight enough to a flying pigeon, or outside the window of a car, that they could survive flight conditions. All lines of evidence provide support for the idea that crayfish may be able to hitchhike thanks to birds. Although these conditions are probably rare, there is a possibility that humans may not be the only culprits when it comes to spreading crayfish outside of their native range.

Solving an invasive species problem by introducing another invasive species is the definition of poor planning. It's like trying to put out a fire by pouring gasoline on it.

Another route through which crayfish can be spread is the education system. Plenty of biological supply companies offer live crayfish for sale. But most of the crayfish offered are species with high invasive potential, which are being shipped around the States. Some classes keep crayfish in aquariums throughout the school year, which is a great way to learn about these animals. But sometimes, a teacher or students may want to release them into a local body of water, thinking they are returning them to their natural habitat. And so educators and students in classrooms in North America may be unknowingly unleashing invasive crayfish into the waterways. It just goes to show that even the most well-meaning among us must be careful not to inadvertently cause more harm than good.

CRAYFISH HAVE BEEN ABLE to travel around the United States, and the globe, due to the human (and potentially avian) activities described above. Their transportation comes with an economic incentive due to the inherent value of these animals, whether through the pet trade, fishing, or aquaculture. But the economic pulse that crayfish can provide is unlikely to outweigh their negative ecological cost. Knowing how crayfish move around is just the first part of the story, because the real damage is not in their introduction but in the disasters that occur once they take hold in a foreign environment.

## A Competitive Advantage

Back to rock-flipping in the pristine yet highly invaded caldera of Crater Lake. There, biologists are on the hunt for one native amphibian that calls this lake its home, the Mazama Newt. This newt is a subspecies of the Rough-Skinned Newt (*Taricha granulosa*), which ranges throughout the Pacific Northwest. But the Mazama Newt resides in one spot and one spot only.

One tiny, four-fingered footstep at a time, Mazama Newts made it into the recently formed caldera a few thousand years ago, and at one point, these not so slimy amphibians were one of the top aquatic predators along the shorelines of this isolated wonder of a lake. But then humans arrived, with their grand plans and reckless interventions.

In the name of recreation, seven species of fish were introduced to Crater Lake in hopes of attracting fishermen. And why stop there? Why not throw in the Signal Crayfish as well, just to spice things up? The fish have to eat, after all. During all this, no one seemed to consider the impact that these newcomers would have on the delicate balance of the lake's ecosystem. The poor Mazama Newts were low on the priority list that day.

The success of Signal Crayfish was not instant, but their dominance has been ramping up in recent years. Roughly half of Crater Lake's shoreline in 2008 was occupied by crayfish. And the unoccupied remaining half had the best spots to find Mazama Newts. Recent numbers paint a bleak picture, as 2020 surveys suggested that the crayfish's takeover is worsening, with 80% of the shoreline now occupied by crayfish. The newts are being squeezed from all angles with nowhere to run. The ultimate fear is complete elimination of the Mazama Newt.

Where crayfish and newts coexist, it's a losing battle for the newts. The newts live their adult lives on the shoreline, where they feed on insects, and they sometimes wander into the water for the occasional aquatic insect. To reproduce, they need to lay their eggs in the shallows. Before the introduction of Signal Crayfish, Mazama Newts had no issues wandering down to the shoreline and dropping off their eggs in the water. But now, instead of just dropping off their soon-to-be offspring, they drop off a snack for the bustling population of crayfish.

Amphibian eggs are full of nutrients and are an easy meal for crayfish. Defending their eggs from the Signal Crayfish that wield claws as large as the entire newt is not an option. The competition and egg predation from Signal Crayfish slowly took its toll on the Mazama Newt population. There is still some hope, as isolated populations of Mazama Newts that are not overrun by crayfish exist, but they are few and far between.

The Mazama Newt is not alone in its hardships. The truth is, the number of species that have suffered and are still suffering a similar fate because of invasive crayfish is unthinkably vast. Pick a random state in the United States, or a random country in the world, and odds are you'll find evidence of destruction from invasive crayfishes. I chose to focus on the Mazama Newt, but I could have chosen among hundreds of species known to have declined in the presence of invasive crayfish. The average crayfish biologist is a cheerful and positive person. But many can bring doom and gloom to any conversation when discussing crayfish invasions.

CRAYFISH ARE THE PERFECT INVADERS. Crayfish aren't too picky and will eat more or less anything. This means that if they get transported from their native bayou in Louisiana all the way to the urban waterways of Los Angeles, they will find food. Transport that same species to an urban pond in Arkansas, and they will find food. Wherever invasive crayfish have been moved to,

Rusty Crayfish (*Faxonius rusticus*) underwater. Photo by Nate Engbretson.

biologists have documented drastic declines in native plants, native aquatic invertebrates like snails and caddisflies, and other aquatic-bound animals like amphibians, which lay their eggs underwater.

Wisconsin is one state with a history of crayfish invasions. One Wisconsin lake in particular, Trout Lake, is well studied for the long-term effects of crayfish invasions. It took nineteen years after the initial introduction of Rusty Crayfish for them to spread throughout the lake, at a rate of 0.68 kilometers per year. Before the crayfish took hold of the benthos, there was an abundance of wildlife exploring the spaces between and underneath the rocks. But afterward, the landscape was dominated by the crayfish. Native snails in the lake alone saw perhaps the biggest decline, from 10,000 snails per square meter to less than five snails per square meter. On top of this, aquatic plant beds decreased by up to 80% in some areas. Findings like this have been reported in dozens of areas around the world.

Rusty Crayfish have spread to numerous lakes in northern Wisconsin, likely due to the popularity of recreational fishing in the area. These individual lakes provide a unique opportunity for biologists to study the impact that invasive crayfish have on ecosystems. One trend that has been observed is the "boom and bust" nature of invasive crayfish populations. After being introduced, crayfish populations explode and dominate the ecosystem, leading to a significant reduction in food sources, including aquatic plants. However, this boom can also lead to the destruction of the crayfish's habitat, because many seek shelter in dense plant growths, causing a decline in the population over time. But the initial boom often drastically impacts the ecosystem and places the crayfish in a spot where they can hold on forever.

I have stumbled upon streams that have recently had Rusty Crayfish introduced to them in central Pennsylvania near the capital city of Harrisburg. In these newly invaded sites, a single step into the water sent hundreds of crayfish scrambling in all directions, tail-flipping furiously like panicked little rowboats. After checking records of invasive crayfish in Pennsylvania, I confirmed that this was a newly established population, which explained the lack of native species, which are typically hidden under the largest rocks. But they were nowhere to be found.

INVASIVE CRAYFISH not only impact plants and other animals, but they also pose a significant threat to native crayfish worldwide. Invasive crayfish can eat up all the resources that were once the property of the native species, outcompete them in fights, and outbreed them by reaching reproductive maturity faster and producing more offspring. If that's not enough, they can also hybridize with native species and dilute the native stock. I flesh out

· · · · · · · · · · · · · · · · · · · · · · · · · · · · · · · · · · · · · · · · · · · · · · · · · · · · · · · · · · · ·

# CRAYFISH PLAGUE

Crayfish plague is a suite of symptoms caused by the fungus-like water mold *Aphanomyces astaci*. North American crayfish have immunity to this pathogen. But for the remaining crayfish around the world, especially in Europe, the same cannot be said. No native European crayfishes have an immunity to crayfish plague. When North American crayfish started to get shipped to Europe (primarily for aquaculture), they started spreading the plague, which has wiped out an estimated 90% of the native European crayfish populations. In most cases, the fact that crayfish plague has reached a population is only noticed when a massive drop in crayfish abundance is observed. In the terminal stages, crayfish rise up, "walk on stilts," and lack their tail-flip response. Eventually, the loss of eyestalks and other limbs may occur. A crayfish plague infection is a death sentence for any susceptible crayfish.

· · · · · · · · · · · · · · · · · · · · · · · · · · · · · · · · · · · · · · · · · · · · · · · · · · · · · · · · · · · ·

these effects in chapter 7, which documents the decline of one of the rarest crayfish in the United States.

On top of the invasive-versus-native competition, invasive crayfishes can spread viral, bacterial, and fungal diseases around with them. Crayfish can serve as a reservoir for *Batrochochytrium dendrobatidis*, or chytrid fungus, and *Aphanomyces astaci*, which causes crayfish plague, which are the world's number one killers of native amphibian and crayfish populations, respectively.

Bad things happen whenever crayfish are spread around and introduced into faraway environments. Competition, disease spread, overconsumption of resources, you name it. But as if all that weren't enough to make us shake in our boots, there is a new threat to crayfish and ecosystems worldwide that is even more menacing. Its invasion potential is greater than that of any other species, and it has the potential to cause untold damage to our delicate ecosystems. And it takes only a single individual to start the destruction.

## Attack of the Clones

In the mid-1990s, rumors began to circulate about a remarkable crayfish found in the German pet trade that could reproduce without mating. This elusive creature was said to be capable of giving virgin birth to up to 700 offspring despite never interacting with another crayfish. Dubbed the Marmorkrebs ("marbled crab" in German), this species was reportedly entirely female—no males were ever found or reported.

At first, many dismissed these rumors as wild exaggerations or flat-out lies. But as word of the Marmorkrebs spread throughout the industry, it eventually made its way to the Crust-L email list—a hub for crustacean enthusiasts to stay informed about all things crustacean. Despite the initial skepticism, some biologists were sent specimens of the alleged self-cloning crayfish. The biologists initially scoffed at the senders, suggesting that they needed to learn how to properly identify crayfish and how to sex them, as every specimen received was female. At the time, it was not thought that a crayfish species could exist if it was entirely female. The experts were so convinced that the Marmorkrebs was a hoax perpetrated by amateur aquarists that they even offered multi-thousand-dollar rewards for anyone who could locate a male.

A few years later, the Marmorkrebs eventually caught the attention of a few other crayfish biologists who were able to get their hands on some specimens. These astacologists took the science-fiction-like claims of virgin birth more seriously. And in 2003, it was confirmed that this crayfish did indeed reproduce asexually without ever mating. This process, known as parthenogenesis, made it among the only known crustaceans to engage in this self-replicating process. It was also confirmed that this species was entirely female, with no males existing.

Later, genetic evidence was gathered that showed that the Marmorkrebs was eerily similar to a rather unassuming crayfish from Florida and Georgia known as the Slough Crayfish (*Procambarus fallax*). But Slough Crayfish are not known to reproduce parthenogenetically, as they undergo typical sexual reproduction like all other crayfishes.

The Marmorkrebs is unlike anything crayfish biologists have ever seen. Out of the blue, starting with rumors in the pet trade and eventually finding itself in the scientific limelight, a crayfish unlocked the ability to clone itself. The story continued to unfold over the coming years. Now there is strong evidence that somewhere, likely within the confines of an aquarium and not in the wild, a typical mating occurred between two Slough Crayfish. Under the usual circumstances, each crayfish parent provides a pair of their chromosomes in either the egg or the sperm, but in this freak mutation event, one parent may have given both pairs of chromosomes to the off-spring due to a mutated sex cell. Marmorkrebs have three sets of ninety-two

A Marbled Crayfish (*Procambarus virginalis*). Wikimedia Commons.

chromosomes, for a grand total of 276. Slough Crayfish, the species the Marmorkrebs derived from, only has two sets of ninety-two. This chromosomal abnormality would typically have resulted in an aborted offspring, but for unknown reasons, the original offspring was able to survive.

In 2017, the Marmorkrebs was officially recognized as being different enough from the Slough Crayfish to be split off into its own species, the Marbled Crayfish (*Procambarus virginalis*). It takes its species name from its ability to give virgin birth. At the time, the paper that described *P. virginalis* ended with a warning message along the lines of "We don't want it to get out." But it was too late. At the time, this species was already found around the globe, not only in aquariums but in the wild.

AS IF THE STANDARD invasive potential of crayfish was not enough, now there is a mutant self-cloning crayfish being sold around the world. In a typical scenario, when a crayfish species gets introduced into an environment, it takes a fair number of males and females to get things going. Or at the very least, a female with some viable eggs. But not the Marmorkrebs, oh no. A single individual can be released, and before you know it, there are millions of them. Now, you might think that a species made up entirely of females wouldn't be much of a threat. After all, where's the aggression going to come from with no males in the picture? But all data suggests that these females are fierce competitors. The Marbled Crayfish has all the makings of a perfect invader.

Soon after news started to get out in the early 2000s about the official discovery of the Marmorkrebs, the crayfish was spreading around the world. Most countries in Europe have introduced populations of this species. And every few years new countries are added to the list. It doesn't stop at Europe, as introductions have occurred in areas as remote as Madagascar, where now Marmorkrebs are the dominant freshwater invertebrate throughout the entire island.

When I initially wrote this chapter in December 2022, no Marbled Crayfish were known to have been introduced in the United States. But since then it has been confirmed that Marbled Crayfish have established themselves in at least one location in Burlington, Ontario, Canada, as well as in at least two locations in New York. And by the time this book is published, it is likely that several more locations will have been discovered, painting a tragic picture for the future. At this point, it is inevitable that Marbled Crayfish will eventually take over environments across the United States just as they have in other countries.

# Controlling the Crayfish

With nothing but doom and gloom in this chapter, I can end on a brighter note by discussing the ways in which the spread of invasive crayfish can be prevented and stopped. But sadly, even this will not be the happy ending you may be hoping for. There are no silver bullets to solve these problems. If anyone figures out how to efficiently get rid of invasive crayfish without killing every other aquatic invertebrate and therefore collapsing the ecosystem, you will be the richest person to ever work with crayfish.

When a crayfish species gets introduced, it is already too late and the water body will never be the same. But don't go thinking that crayfish biologists are just sitting around feeling sorry for themselves. Hundreds of scientific papers document the successes and failures of numerous methods of removal. Some of these methods are intuitive, whereas others are unconventional.

Setting out traps may be the most obvious solution but is unfortunately not that easy. Yes, luring them in and hauling them out can certainly make a dent in the invasive population, but don't be too quick to break out the celebratory butter and Cajun seasoning just yet. Routine trapping can help reduce the population, but it's unlikely to result in full-fledged removal.

A crayfish trap attempting, unsuccessfully, to capture Rusty Crayfish (*Faxonius rusticus*). Photo by Nate Engbretson.

Baiting the traps with the smelliest, grossest bait you can find may help, but it still only catches the most active crayfish—typically larger males out for a stroll. Juveniles and egg-carrying females tend to hunker down, making them much harder to trap. As a result, even if you remove a majority of the males via traps, the population has a stockpile of females, which keeps their numbers high throughout the year. Even studies conducted on small ponds with daily or weekly trapping have shown minimal long-term effects on eradication. It's like being in a never-ending battle: No matter how many crayfish you catch, there always seem to be more hiding just out of reach.

Several studies suggest that combining eradication methods is required to really make a dent. Limiting the number of sport fish that humans catch, like rock bass, can be one way to decrease crayfish numbers. When fewer predatory fish are caught, they will eat more crayfish.

Some researchers have even used sex pheromone lures, where a gel mixed with female crayfish sex pheromones, or just an adult female, is added to entice sex-driven males into the trap. Similar pheromone lures that bring in males during the breeding season have been used widely to contain insect pests but seem to be less effective in crayfish.

A large-mouth bass playing with its food while trying not to get pinched.
Photo by Nate Engbretson.

Instead of trapping, some researchers have turned to a technique known as sterile male release technique (SMRT). This approach aims to reduce the population by making the crayfish sterile. Here's how it works: Male crayfish are naturally promiscuous and aggressive during the mating season, often seeking out multiple mates. Even a male on the lower end of the size scale will mate with at least one female, while a big-clawed beast might manage to mate with up to six. Then, a male delivers his sperm packets to the female(s), who then carries the fertilized eggs until they hatch into baby crayfish. But what if the male were sterile? If a male crayfish mated rampantly, the female would still lay her eggs, but they wouldn't be viable. And that's the idea behind SMRT: By releasing large numbers of sterile male crayfish into the population, researchers hope to dramatically reduce the number of viable eggs being produced.

Sterilization is typically done through exposure to radiation, which fries the males' reproductive system. In some insects, a single dose of radiation can result in individuals that are perfectly normal except that 90% of their sex cells are sterile, resulting in large-scale decreases in invasive populations. These methods have not been fully worked out for crayfish though, and studies report only a 40% reduction in reproductive output. Future work in this realm may be one of the best options to combat invasive crayfish populations.

AS MUCH AS HUMANS love a success story, the truth is that complete removal of invasive crayfish populations is a rare feat. In fact, it often requires some drastic measures that are only feasible in small and isolated areas, like a tiny pond or canal. If draining a pond is possible, this is your best course of action. But draining and hand-picking the flopping crustaceans at the bottom is not always going to be enough, because depending on the species, they may simply hunker down in a burrow and wait for the pond to fill back up. Invaders like the Red Swamp Crayfish are known burrowers, and they can hide out in their burrow for years.

Even if you can drain the water, there's still no guarantee that the crayfish will be gone for good. In one case, a dam in South Africa was drained and all the crayfish were physically removed, but twenty-eight years later, the Red Swamp Crayfish had returned! These critters are just too darn adaptable, which is what makes them such successful invaders in the first place.

If you can't drain the pond, you might have to resort to chemical warfare. But as with any war, collateral damage is a risk, as chemicals can harm all sorts of creatures, not just the crayfish. One popular chemical option, Rotenone, has proven to be effective, but the concentrations needed to kill

# CRAYFISH BANS

To combat the spread of invasive crayfish from negligent owners, several states have enacted bans on the buying, selling, or owning of crayfish. For example, in Pennsylvania, a 2015 law makes it illegal to own any crayfish except when they are sold for scientific research or for restaurant consumption. This means that selling live crayfish bait is illegal in Pennsylvania, as the fishing industry is presumably the source of most aquatic invaders in the state.

Similar bans are starting to be passed in other states that are being overrun by invasive crayfish. And every few years you hear news stories about crayfish breeders getting busted. Many of these laws are now focusing on preventing the Marbled Crayfish (*Procambarus virginalis*) from gaining a foothold in the States, as places like Michigan and Ohio have specifically banned the ownership and sale of the Marbled Crayfish.

crayfish are forty times what it takes to kill a fish, which is obviously a concern. Fortunately, there are other options, such as Pyblast, which is less toxic to mammals and birds and doesn't stick around in the environment for too long. In an ideal world, a chemical could be created to impact only the invasive crayfish and leave native crayfish and other animals alone, but hoping for such a treatment may be blind optimism.

ONCE INVASIVE CRAYFISH are established in an area, it takes everything short of a miracle to get them out. Control and population reduction methods aren't good enough yet to confidently remove a population of invaders. Although eradication programs—which state and federal agencies pump money into—are necessary for management, many believe that the best way to prevent crayfish from spreading is to focus on preventing the introduction from occurring in the first place.

"An ounce of prevention is worth a pound of cure" is the motto of any conservation educator when it comes to invasive species. Once an invasive species takes hold after an initial introduction, there is little to no hope of removal. Therefore, instead of funneling millions of dollars into removal programs, such as trapping the same lake every day for years with no hope, preventative and educational approaches can be taken.

Whether invasive species introductions are intentional or accidental, education is becoming a favored tool for managing this problem. In areas where the use of crayfish as bait is widespread, targeted outreach to local fishermen and bait shops can go a long way in reducing their spread. Signs, discussions, and educational campaigns are all proven ways of keeping the public informed and engaged in the fight against invasive crayfish.

Many experts believe that funding for new programs should prioritize prevention rather than removal. By reducing the pathways of risk and establishing early detection systems, management officials can stay ahead of the problem. Of course, the issue is far from simple. Different species, habitats, and scenarios require unique approaches, and this requires adaptation from the management standpoint. One thing is certain, if nothing is done, invasive crayfish will continue to spread and wreak havoc.

THE SPREAD OF CRAYFISH, the effects of their establishment, and the strategies for preventing and managing these invaders has been thoroughly documented. The next chapter will provide a closer look at an example that demonstrates just how troublesome the invasive crayfish problem can be. And it's not in the heart of a bustling city but rather in a remote region far from a major metropolitan area. Despite its isolation, this area is not immune to the devastating effects of invasive crayfish, which threaten to cause the extinction of one of the rarest crayfish on the planet.

# CHAPTER 7

# *Pacifastacus*

.........

## THE CRAYFISHES OF THE
## PACIFIC NORTHWEST

### Winter Snorkeling for the Shasta Crayfish

DESPITE THE FRESH three inches of March snow on the ground, I am preparing to drop into the outflow of a chilling spring-fed lake dressed in a dry suit and snorkel. By my side is Maria Ellis, who has partaken in hundreds of such cold-water snorkel surveys over the past thirty-plus years. Beforehand, Maria tells me that this lake has not been searched for crayfish since the year I was born; issues with the affluent landowners of this private property have prevented Maria and her husband Jeff Cook (whose dry suit and snorkel I am thankfully borrowing) from setting foot on this expansive lakeside property in Northern California. (At Maria and Jeff's request, I'm keeping information about this location, as well as other locations in this chapter, vague. Although their work has no effect on private landowner rights, the public's perception of conservationists is not strong in this area. Maria and Jeff have sometimes experienced hostility toward their efforts and received viable threats to foil their crayfish conservation efforts.)

On the dock nearby, Maria, Jeff, and I observe several piles of otter scat littered with bits of carapaces, claws, and antennae. This confirms our hunch: There are crayfish here. But based on the features of these body parts, it is clear that these bits and pieces are not from the crayfish we are searching for; they are from the Signal Crayfish (*Pacifastacus leniusculus*), a species that did not occur in this area the last time Maria and Jeff snorkeled here. The invasive Signal Crayfish, once introduced into an area, has no rules or limitations to prevent it from spreading.

Even though the landowners had permitted Maria's and Jeff's surveys earlier in their careers, the landowners' mindsets changed and they decided that surveying and monitoring the status of an endangered species on their land was not high on their priority list. Such hesitancy is common throughout this section of Shasta County, California, where landowner permission is required to snorkel. But now, half of this spring-fed lake and the river that flows out of it were recently donated to the Nature Conservancy, a nonprofit environmental organization. The other half is still privately owned and remains off-limits. Maria, Jeff, and I are searching for the endangered Shasta Crayfish (*Pacifastacus fortis*) in these waters for the first time in twenty-seven years. But with the invasive Signal Crayfish now having moved into the Shasta's native territory, Maria, Jeff, and the rest of their team of environmental and wildlife biologists at Spring Rivers Ecological Sciences are not sure whether the endangered crayfish would have been able to hang on; other similarly invaded populations have disappeared completely in a shorter time frame.

AS WE KNEEL OUR WAY down face-first into the water with our snorkels pointed in the air, my exposed face becomes aware of the cold. The dry suits combined with two layers of long pants and three layers of long-sleeved shirts (which I also borrowed from Jeff) will fend off frostbite for at least an hour. Snorkeling for crayfish in early March with snow on the ground may seem confusing, but the temperature of the water, a chilling 11°C (51.8°F), that flows out of the groundwater springs here fluctuates very little throughout the year. Snorkeling is a year-round activity for crayfish sampling in this part of the country because the water temperature flowing from the springs varies by less than one degree Celsius from the coldest day in the winter to the hottest day in the summer. And because these waters are spring fed, the water depths are at an ideal height for snorkeling—with rock-flipping only an arm's length away. Although I was told that this water has among the lowest visibility of the spring-fed systems in the area, the water appears crystal clear to someone like me, who is used to the naturally muddy, tannin-soaked brown water found throughout Appalachia.

Now completely immersed, Maria drifts off to flip rocks in her own direction, while I am awkwardly floating on top of the water trying to get my bearings. Maria seems just as comfortable snorkeling as she does out of the water, as she has dedicated her life to protecting Shasta Crayfish and other animals in this area. She started this work during her PhD research at the University of Michigan and never had the heart to leave the Shasta Crayfish. She knew that without a dedicated conservationist working directly with them, they would not last. This led her to start Spring Rivers, which comprises both an ecological consulting company and a nonprofit. Maria met Jeff, her husband and cofounder of Spring Rivers, when he started working in the area while she did her PhD, and both have now spent most of their lives working with the Shasta Crayfish.

With my gloved hands, I pull back some rocks that vary in size from something I can palm to boulders the size of a large serving plate. The rocks here are all lava rocks, courtesy of the extensive basalt lava flows that formed the Modoc Plateau, from which this water flows. Also nearby is Mount Lassen—home of Lassen Volcanic National Park—which I have noticed looming large on the horizon on several occasions during my time in the area. The lava rocks that make up the substrate here are all porous, gray, and covered with small air bubble indentations. This unique environment, filled with natural springs, looming mountains, and volcanic bluffs, is typical of this area, located at the junction of three physiogeographic regions: the Sierra Nevada Mountain Range, Cascade Range, and Modoc Plateau.

With each flip, I notice that the underside of each rock is caked in a layer of freshwater sponges—filter-feeding animals that live mostly in marine environments, but some occur here in the cleanest of freshwater. And on the rocks, under the rocks, and scattered throughout the environment are pea-sized snails in the genus *Fluminicula*. Strangely, when I lift a lava rock, instead of seeing the sandy lava substrate, all I am finding is a one-inch-thick layer of *Fluminicula*, with dozens of small freshwater isopods and amphipods (both small crustaceans) squirming and squeezing in among the snails. The chilling constant temperature and crystal-clear water paired with this strange community of animals feels like a different planet that deserves the attention of a David Attenborough–narrated documentary.

I finally get acclimated to this newfound underwater planet and start to carefully flip some of the larger rocks—hoping to find some larger

A typical view while snorkeling in one of the many springs scattered throughout Shasta County, California. Photo by Koen Breedveld.

crustaceans. When I peer under the largest rock I have flipped so far, a moderate-sized crayfish darts out, exhibiting the stereotypical tail-flip escape response. After a clumsy snatch with my nearly frozen gloved hands, I end up grabbing this crayfish out of the water mid-escape. Without even looking at the crayfish, just based on its escape attempt, I know that this is not the crayfish I am looking for. And after a closer underwater examination, I confirm my hunch. The crayfish's behavior in my hand and its body shape, coloration, and rostrum (the extension of the carapace between the eyes) all indicate that this crayfish is the villain in this environment: the invasive Signal Crayfish.

Although a subspecies of the Signal Crayfish is native in the northwestern corner of California, some 200 miles away in Shasta County in the northeastern corner of the state, and in the rest of California, Signal Crayfish are highly invasive. Since their initial introduction to the area by fishermen sometime in the 1970s, they have invaded this waterway and taken over most of the Shasta Crayfish's habitat. The last time that Maria and Jeff sampled this secluded spring, there were no Signal Crayfish, but there were Shasta Crayfish. But Maria and Jeff knew that the spread of Signal Crayfish was inevitable and that they would soon take over, just as they had witnessed happening at so many other locations in the last three decades.

With the invasive Signal Crayfish in hand, I snorkel my way over to Maria to show her my find. I hear her murmur "Signal" underwater with the snorkel's mouthpiece still in, as she continues flipping rocks. With her confirmation, I slide the crayfish into one of the two custom-made underwater containers Maria has with her, which allow crayfish to be placed inside but not escape. My crayfish joins the six Signals Maria has already collected. Maria's continued rock-flipping and unwillingness to stop for even a second are a testament to her worry that we will find nothing but Signals and that the Shasta Crayfish have vanished from yet another location.

IN HOPES OF FINDING A SHASTA, I move downstream to a patch of larger boulders. Underneath every rock is the same strange allotment of sponges, isopods, and amphipods—but no crayfish. While I'm using both of my hands to lift a nearly ten-pound boulder, I hear an exclamatory mumble from Maria in the distance. With the water nearly the exact depth as the length of my arms, I hurriedly "knuckle drag" my way over to Maria, as she pops off the mouthpiece of her snorkel and exclaims, "Shasta!"

In Maria's hand is one of the rarest crayfish in the world. This crayfish is hunkered down and relaxed in her open palm, which I learn is typical for a Shasta. Although this Shasta Crayfish is related to the Signal Crayfish (they both belong to the genus *Pacifastacus*), Shasta Crayfish share only a scant resemblance to their cousins. The animal in Maria's gloved hand is a color that is difficult to describe, with a primary coat of grays, blues, and purples, depending on the angle that you look at it, with dozens of small white dimples covering their entire exoskeleton. On the ventral side of the exoskeleton is a fluorescent orange, which can creep its way onto the edges of the carapace and tops of the claws. Shasta Crayfish have a robust, stout body, with stocky missile-shaped claws that look like they could draw blood with no issue—but after years of working with these animals, Maria claims that getting pinched by a Shasta is a rare occurrence. One of the first common

Most Shasta Crayfish (*Pacifastacus fortis*) have a coloration similar to the lava rocks they live under. Occasionally, Shasta Crayfish are found with a blue-tinted carapace (*bottom*) and less vibrant, more pinkish color under their claws. Photos by Koen Breedveld.

names for *P. fortis* was the Placid Crayfish, relating to its calm and peaceful demeanor. The official common name, however, is the Shasta Crayfish, as it only occurs in Shasta County, California.

This cool, calm, and collected crayfish in Maria's hand is roughly 3.5 inches (almost 9 cm) long, which, based on the crayfish that I have worked with, equates to around a two- or three-year-old. But Shasta Crayfish don't follow the same rules that I have come to know, and the data Maria has collected suggests that this specific animal is in the ballpark of five or six years old—just reaching the age of reproductive maturity for this species. The largest Shastas, roughly double the size of the one in Maria's hand—are estimated to be at least ten to twelve years old.

Although it behaves as if there is nothing to be afraid of, lying in Maria's hand, this species is on the verge of extinction. There is a justified fear that Shasta Crayfish will go extinct within my lifetime.

AFTER STARING AT THIS CRAYFISH and repeatedly noting its beauty (that is what you do when you are a crayfish nerd), Maria and I are both energized in hopes of finding more Shastas. Maria places the male she caught into its own underwater container (making sure to separate it from the half dozen Signals that we already caught), where it will rest until we are done snorkeling and can measure its size and record additional data.

With newfound energy, I am back to flipping rocks, completely ignoring the fact that I can't feel my numbing fingers and lips. Now, more agile in the water than I was twenty minutes ago, I wander to a prime-looking patch of lava boulders. Each of them is barely embedded into the smaller cobble underneath, which makes them an easy-access shelter for Shastas. After lifting a boulder the size of a laptop, I see the usual suspects that are underneath every rock. But tucked away amid a layer of snails is the dark body of a Shasta Crayfish. Recognizing this crayfish takes me a second, because not only is it buried under a layer of snails, but it is also missing both of its claws. Based on the ratio of Signal Crayfish to Shasta Crayfish that we have found so far (and the marked behavioral differences between the aggressive Signals and the placid Shastas), it isn't hard to imagine that this clawless Shasta had recently gotten into a scuffle with a much larger Signal Crayfish, which can reach nearly double the size of a Shasta in only a quarter of the time.

Over the next half hour, Maria and I collect roughly forty crayfish, but only five of these are Shasta Crayfish; the rest are the invasive Signal Crayfish. This is the reality that Maria and her husband Jeff are now facing. The last time Maria sampled here, she found a decent-sized population of Shasta Crayfish. But she knew it was only a matter of time before the Signal Crayfish spread upstream to invade the Shastas' native habitat. For now, the Shasta Crayfish are hanging on in their competition with the Signals, but they are not the favorites in this contest.

Out of the dozen naturally occurring populations of Shasta Crayfish that remain, every single one of them is now living with Signal Crayfish—their invasive, pesky cousins who outcompete them at every stage of life. In some locations, the Shastas have been able to hang on, albeit in lower numbers, whereas other populations disappear almost immediately after invasion. Thirty-plus years of snorkeling and removing Signal Crayfish by hand has delayed the Shasta decline. But these removal techniques are just a Band-Aid that constantly needs to be reapplied: Efforts to completely removing the invasive from a location have never been successful. For this reason, the Shasta Crayfish is one of the most imperiled freshwater species in North America.

# Across the Continental Divide

The native Pacific Northwest (PNW) crayfish are a unique bunch and the only members of the taxonomic family Astacidae in North America. Although there's no universally accepted definition of the PNW, for my purposes I consider it the northwesternmost part of the United States, which includes Washington, Oregon, Northern California, and Idaho. With all eight species in the genus *Pacifastacus* being from the PNW, their stories encompass nearly every situation in modern crayfish biology: One is a worldwide invader, one is on the brink of extinction, another has already gone extinct because of humans, we know almost nothing about four of them, and the last one is known only from fossils.

As we've seen, the Shasta Crayfish (*Pacifastacus fortis*) is critically endangered and is being replaced by a worldwide invader, the Signal Crayfish (*P. leniusculus*). The Sooty Crayfish (*P. nigrescens*) has not been collected since the 1870s and is presumed extinct. It is known only from three preserved specimens in the Smithsonian National Museum of Natural History (NMNH) in Washington, DC—the rest of preserved Sooty Crayfish specimens were destroyed in the Great Chicago Fire of 1871. The last four extant PNW crayfishes, the Pilose Crayfish (*P. gambelii*) the Snake River Pilose Crayfish (*P. connectens*), the Misfortunate Crayfish (*P. malheurensis*), and the Okanagan Crayfish (*P. okanagensis*) are extremely understudied. Little is known about their biology outside of the fact that they are also being impacted by at least three invasive crayfishes. With the exception of the Signal Crayfish, the members of the genus *Pacifastacus* are having hard times.

# FOSSIL CRAYFISH

One of the eight PNW crayfish, *Pacifastacus chenoderma*, is an extinct species from the Green River formation in modern-day western Idaho and eastern Oregon. Geologic estimates from the rock formations suggest that this species has been extinct for about 3–4 million years. *Pacifastacus chenoderma* reached large sizes—comparable to the impressive body lengths reached by Signal Crayfish (*Pacifastacus leniusculus*), with well-preserved carapace fossils 60 millimeters (2.4 in) in length.

The hard exoskeleton of crayfish, combined with the propensity of some species to create burrows, lends itself well to the fossilization process. Throughout the world, there are at least a dozen examples of fossil crayfish, with new species discovered every few years.

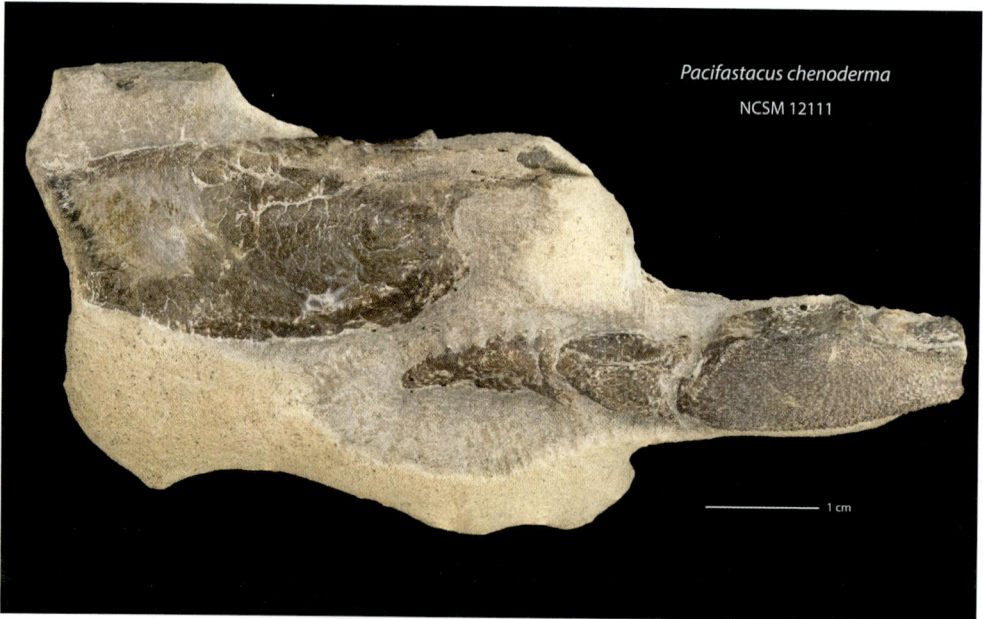

*Pacifastacus chenoderma*
NCSM 12111

1 cm

Side view of a *Pacifastacus chenoderma* fossil currently housed alongside several other specimens in the North Carolina Science Museum. Photo by Bronwyn Williams.

Not only are the crayfishes in the PNW facing unique issues, but they also have aspects of their biology that separate them from all other North American crayfishes. *Pacifastacus* crayfishes belong to the taxonomic family of crayfishes known as the Astacidae (often referred to as the Astacids). East of the Rocky Mountains, across the Continental Divide, every crayfish species belongs to a different taxonomic family: the Cambaridae (often referred to as the Cambarids). Although both are crayfish, the most recent common ancestor of these two families is estimated to have existed around 150 million years ago! And since then, the *Pacifastacus* crayfishes evolved different body shapes, different lifestyles, and different mating habits. In many ways, these historical adaptations are now inhibiting their ability to fend off the more generalist invasive competitors.

EARLY CRAYFISH BIOLOGISTS realized that the crayfishes in the PNW weren't all that similar to the crayfishes found in the Eastern United States. For one, female Astacids lack an *annulus ventralis*—the sperm-storage system that female Cambarid crayfishes have. With females lacking an internal sperm-storage unit, male Astacids deposit spermatophores directly onto the underside of the female when they mate. Male Astacids also lack reproductive form alteration. That is, they can reproduce year-round and do not alternate between being in reproductive and nonreproductive forms like Cambarids do.

Because of these differences, early scientists initially placed the PNW crayfishes in the genus *Astacus*, which is a group of crayfishes from far, far away: Europe and Eurasia. Later, when the PNW crayfishes were placed in their own genus, *Pacifastacus* (i.e., Pacific *Astacus*), it was still widely held that, oddly, they resembled the European *Astacus* crayfishes that lived 5,000 miles away. Modern genetic tools helped confirm this early hunch: The crayfishes in the PNW are most closely related to European crayfishes and not the crayfishes in eastern North America. But how does this make sense?

To unravel these biogeographical (i.e., relating to the geographic patterns of distributions and diversity of life) puzzles, you need to look back in time and take a historical and geologic glimpse at earth's history. Modern earth is separated into seven continents and thousands of islands—all of which are relatively new developments when compared to the age of the earth. But around 225 million years ago, geography class would have been easier, because there was only a single supercontinent, known as Pangea. Without having to worry about oceans getting in their way, animals and plants could spread wherever the climate and local ecology suited them. There is strong evidence that crayfish were around at this time, likely doing all of the same things they do today: digging burrows, fighting with each other, eating anything they could get their hands on, and being eaten by anything that can get their hands on them.

Slowly but surely, due to the shifting of the subsurface plates on which the earth's crust rides, the supercontinent of Pangea split into two smaller supercontinents, Laurasia in the Northern Hemisphere, and Gondwanaland in the Southern Hemisphere. When Pangea separated, crayfishes (and many other forms of life) were also separated, creating a major divide in the evolutionary trajectories of these animals.

The chunk of land that was Laurasia comprised North America, Europe, and Asia. During this time, what is now New York City was not a ten-hour plane ride away from any modern European city but more so a short drive. And with no oceans to divide these ancient crayfishes, the ancestor of both the PNW *Pacifastacus* and the European *Astacus* was evolving in Laurasia. Fast-forward another 100 million years and plate tectonics had continued to do its thing of spreading apart these landmasses. During this time, the *Pacifastacus* crayfishes settled in what's now the PNW, and their *Astacus* ancestors settled in Europe and Eurasia. So they likely came from a single population living on the northern supercontinent but eventually diverged from each other when Laurasia broke apart.

The PNW *Pacifastacus* and the European *Astacus* crayfishes started to adapt to their respective new environments, but they have dozens of traits in common, which probably led to their biological success in their respective areas early on. But now, due to the presence of humans and the spread of invasive species, those same once-beneficial traits are making the Astacidae family the most imperiled of any group of crayfishes.

Like their European relatives, *Pacifastacus* crayfishes generally reach large body sizes. This is especially true for the Signal Crayfish, whose carapace alone can reach up to 70 millimeters (2.7 in) in length. Other *Pacifastacus* species, like the Shastas and both Pilose crayfishes, are smaller, but they grow at a much slower rate. Signal Crayfish reach their massive sizes and sexual maturity within two or three years. Shasta Crayfish, by comparison, do not reach sexual maturity until they are around five or six years of age.

Not only do Signal Crayfish reach sexual maturity faster, but female Signals also produce over 100, and in many cases up to 200 or more, offspring with each breeding event. Comparatively, a female Shasta Crayfish can produce up to seventy eggs during one breeding event, but finding a female with ten to thirty eggs is typical. Shasta Crayfish grow slowly, reach sexual maturity late in life, and have relatively small batches of offspring compared to the Signal Crayfish. This difference in the pace of life is a major contributor to the decline of Shastas when Signals are around. There is no way for the Shastas to compete with the Signals' reproductive ability. With their high reproductive output, ability to survive in nearly any habitat, and aggressive demeanor, Signal Crayfish are in many ways a perfect invader.

But the story of the Signal Crayfish's invasive prowess and its ability to outcompete other members of the genus *Pacifastacus* sadly has already played out in the past: Signal Crayfish were likely a critical factor in the extinction of another member of the genus, the Sooty Crayfish. Unlike the sheltered Shasta Crayfish, this now-extinct species did not live a secluded lifestyle in headwater springs systems in a remote part of the country. Rather, the Sooty Crayfish inhabited creeks and streams in California near what would become one of the largest cities in the United States.

## The (Presumed) Extinction of the Sooty Crayfish

For the Shasta Crayfish, things aren't looking good—their highly invasive cousins, the Signal Crayfish, are coming in and invading their territory, outcompeting them, replacing them, and most common, wiping them out entirely. Maria Ellis, Jeff Cook, and their team at Spring Rivers have been combating the decline of Shastas for three decades. One by one, Shasta Crayfish populations have declined. The ultimate fear is their extinction.

Looming in the back of Maria's and Jeff's minds is the fact that one of the two crayfish species to have been declared extinct in modern history is the Shasta Crayfish's closest relative: the Sooty Crayfish (*Pacifastacus nigrescens*). It's unlikely that this is just a coincidence, because these animals share traits that make them exceptionally vulnerable to invaders.

*Pacifastacus nigrescens* has a perilous history filled with countless confusions and disasters. The Sooty Crayfish closely resembled the Shasta Crayfish, except the Sooty Crayfish lacked the orange vibrancy on its underside. However, unlike the Shasta Crayfish's relatively secluded distribution, which keeps it far from urbanization, the Sooty Crayfish inhabited streams and creeks on the southern outskirts of San Francisco. Before this crayfish was ever studied, it was already in serious decline. Being in the confines of a city is never a good thing for a crayfish. The combined pressures of urbanization, overexploitation (humans were eating them), and introductions of invasive crayfish likely led to the Sooty Crayfish's extinction. The last occurrence of this crayfish is unknown, but published reports suggest it was last collected in the mid- or late 1800s.

Sooty Crayfish were once common in many of the larger streams around the San Francisco Bay in California, including Alameda Creek in Alameda

County and Coyote Creek in Santa Clara County. Modern aerial looks at these areas reveal almost nothing but highways, strip malls, and parking lots. In addition to being located near San Francisco when the city's population exploded in the mid-1800s, there are several other oddities related to this crayfish's history. The first Sooty Crayfish ever recognized by scientists were not wild caught—biologists purchased them from a fish market in "the vicinity of San Francisco." Because crayfish were an easily accessible source of protein, Gold Rush–era San Franciscans were likely consuming Sooty Crayfish by the thousands.

The only remaining evidence that this species even existed is in three jars at the Smithsonian National Museum of Natural History in Washington,

The only remaining specimens of the extinct Sooty Crayfish (*Pacifastacus nigrescens*). Photo by Zachary J. Loughman.

DC—two males and one female. Additional preserved specimens of *P. nigrescens*, including the type series (the specimens collected and used in a species' original description), used when the species was first described in 1857, were all destroyed in the Great Chicago Fire of 1871. Having held these three jars myself during a visit to the Smithsonian, I can attest that cradling every known specimen of a species in your hands—pickled in ethanol or not—invokes a chilling sense of melancholy. It's an unsettling experience, stirring up not just a haunting feeling but a cascade of mini existential crises, as you confront the weight of what it means to hold every remnant of an entire species in your grasp.

OUR KNOWLEDGE OF THE BIOLOGY of the Sooty Crayfish is nonexistent. Based on morphological similarities, the Sooty Crayfish is considered the closest known relative of the Shasta Crayfish. The Sooty Crayfish was likely a late-maturing, long-lived, low-fecundity species whose behavior may have been similar to that of the Shasta Crayfish. The late maturation and low fecundity, as well as the mild-mannered demeanor, of these two species likely played a large role in their declines. As Maria Ellis says, "Place five Shasta Crayfish in a bucket and five Signal Crayfish in the same bucket, and even someone that has never seen a crayfish could sort them out into two groups": one group of more docile individuals (the Shastas) and another very aggressive group (the Signals).

The California Gold Rush transformed San Francisco from a small settlement of about 200 residents in 1846 to a boomtown of about 36,000 by 1852. Crayfish from creeks surrounding the San Francisco Bay became a popular item in the city's fish markets, and soon overharvesting decimated the Sooty Crayfish populations. In order to meet demand at the fish markets, Signal Crayfish were imported from Oregon and introduced into the streams where Sooty Crayfish lived. Although human consumption and urbanization certainly led to the steep decline of this species, the introduction of invasive Signal Crayfish was probably the final blow for the Sooty Crayfish. Across the geographic range where Sooty Crayfish were once collected, there are now bountiful populations of invasive Signal Crayfish—a story that currently seems to be repeating itself, not only for Shasta Crayfish but also for both of the Pilose Crayfish species in Oregon, Idaho, Utah, Nevada, and Wyoming.

The spread and competitive edge of Signal Crayfish is truly remarkable and multifaceted. Laboratory studies have demonstrated that Signal Crayfish on the invasion front (the expanding range of the introduced species) are much more aggressive than Signal Crayfish in the invasion core (the now established regions of the population). Signal Crayfish are so dominant in these areas that they also outcompete other invasive crayfishes in the area. About a decade before the Signal Crayfish invaded Shasta territory, the Virile Crayfish (Faxonius virilis) was introduced to the area. Although the Virile Crayfish was the first invasive species introduced here, Signal Crayfish replaced them throughout most of the Pit River and relegated the Viriles to small populations by outcompeting them.

Signal Crayfish are also known cannibals, as they can be predators to their own species as well as to any unsuspecting Shastas or Viriles. Maria has observed Signal cannibalism in both the field and laboratory on dozens of occasions, as well as their attempted predation of Shastas in the laboratory (which she always stopped). As a part of her PhD research, Maria studied the competitive dynamics of Shasta Crayfish and Signal Crayfish in both lab and field experiments. She repeatedly found that Signal Crayfish dominated Shasta Crayfish, outcompeting Shastas for food and shelter and preying on them.

Maria also found that male Signal Crayfish would mate with female Shasta Crayfish. Apparently Signal Crayfish will mate with anything that looks like a crayfish. Based on their aggressive nature and the disproportionately high numbers of Signals compared to Shastas in the wild, male Signal Crayfish often mate with female Shasta Crayfish. Although a Signal Crayfish's sperm mixed with a female Shasta's eggs does not create fertile offspring, the female Shasta Crayfish are effectively prevented from producing baby Shasta Crayfish for the year, a problem known as reproductive interference. Because Shasta Crayfish are a slow-growing, late-maturing, low-fecundity species, reproductive interference is particularly harmful. Together with competition and predation, reproductive interference is part of a triple whammy that Signal Crayfish inflict on Shastas.

In addition to the direct threats posed by Signal Crayfish—competition, predation, and greater reproductive abilities—Signal Crayfish and other invasive species can also introduce diseases into the environments they invade. All organisms have a host of microbes and pathogens with which they interact and from which, in most cases, they have acquired some immunity. But when humans move crayfish across typically unpassable barriers, like across the ocean, they can introduce a novel set of pathogens into the new environment.

Based on the speed of decline once Signal Crayfish are introduced into some Shasta Crayfish populations, Maria believes that pathogens and the spread of diseases may also be playing a role in the decline of Shasta Crayfish. There are a handful of instances where previously dense Shasta populations were wiped out within a year once Signal Crayfish were introduced.

Studying the spread of disease is difficult, however, because there are not always easily diagnosable symptoms of disease transmission.

IN 1977, WELL OVER 100 YEARS since invasive Signal Crayfish were introduced in California, Raymond Bouchard declared that the Sooty Crayfish should be considered extinct. Bouchard was a prominent astacologist, having worked with crayfishes from across the United States and describing twelve new species throughout his research career. After suggesting that Sooty Crayfish should be declared extinct, he also suggested that continued searches for *P. nigrescens* were warranted, though finding one was unlikely. Still, every once in a while, an animal presumed extinct is found again. The oddly charismatic Somali elephant shrew, with its long elephant-trunk-like nose comes to mind; having not been seen for fifty years, it was presumed extinct until it was located again in August 2020. Although human consumption of crayfishes has declined in the San Francisco Bay area, the ever-present threats of habitat destruction and the invasive Signal Crayfish make it unlikely that any Sooty Crayfish remain.

Sooty Crayfish were gone and replaced by the dominating Signal Crayfish before biologists even understood what made them unique. Hopefully this situation does not repeat itself with the two Pilose Crayfishes, *Pacifastacus gambelii* and *P. connectens*, which are dealing with three different invasive crayfish species that have the potential to wipe them out. The two Pilose Crayfishes are among the most understudied crayfishes in the United States. They are the closest living relatives of the Shasta Crayfish. If no one starts paying attention to these animals, then their declines may be just as rapid as that of the Shasta Crayfish.

The Sooty Crayfish's story hopefully does not foreshadow what the future holds for Shasta Crayfish (and the Pilose Crayfishes). It is the worst fear of every crayfish biologist: If nothing is done, then these animals will disappear forever, in some situations without anyone knowing, just like the Sooty Crayfish.

## Protecting the Shasta Crayfish

There are dozens of freshwater spring systems in this region of the Modoc Plateau like the one Maria Ellis and I snorkeled in. Two to three decades ago, many of these springs held stable populations of Shasta Crayfish, without any Signal Crayfish. Because of their crystal-clear water, trout populations, and raw beauty, however, most of these springs are at private ranches and vacation getaways.

An adult Shasta Crayfish (*Pacifastacus fortis*) in its natural lava-rock habitat covered with small aquatic snails. The diet of Shasta Crayfish is unknown, although it is presumed that the abundant snail populations that co-occur with Shastas are a primary source of nutrients. Photo by Koen Breedveld.

The springs were originally home to three Pit River tribal bands, the Achumawi, Ilmawi, and Atsugewi. These Indigenous groups were well acquainted with the ecology of the animals that thrived in this habitat. Across the springs, they constructed fish traps that consisted of walls, of lava rocks stacked up above the water surface level and with a narrow opening. Because flows and water surface levels in the springs fluctuate minimally, the rock walls remain above the water level year-round. Sacramento suckers, a native fish, move in large numbers up into the spring headwaters to spawn in late winter when many other food sources run scarce. Suckers naturally seek out areas of shallow water for spawning and congregated in the traps of their own volition. The tribes then trapped the suckers by closing off the narrow opening of the spring.

To snorkel at an isolated spring system in the Fall River Valley, we again have to request landowner permission before setting foot on the expansive property. On driving up to the spring, I am reminded of the repeated stories Maria has told me of snorkeling here for years and "always finding Shasta

Crayfish in the fish trap." I'm not sure exactly what to expect of the fish trap, but I have high hopes for this location. Once we approach the spring, I finally see the esteemed trap, which likely dates back hundreds of years. Fish trapping has not occurred in this sheltered section of the spring for some time. The lava boulders in the fish trap were once home to the most dense and abundant Shasta Crayfish population ever documented, according to Maria. She has recounted stories to me of flipping a single large rock in this small area in the spring and observing more than two dozen Shasta Crayfish underneath—numbers she is now lucky to see in an entire year of collecting. But now, the only thing you will see in this habitat is the tail flip of a fleeing Signal Crayfish.

This specific location was once the last major refuge for the Shasta Crayfish, because until 2005, it was the only area known to have withstood invasion from Signals. Before 2005, Signal Crayfish had invaded every single other body of water in which Shasta Crayfish occurred. It was known, however, that Signal Crayfish were present downstream of the spring and that it was only a matter of time until they made it upstream to this isolated population. As it was the last isolated population of Shastas, Maria and Jeff tried to prevent the upstream migration of Signal Crayfish to the area.

Their goal was to create a barrier that was passable for migratory fish like the suckers and trout but impassable for Signal Crayfish. In collaboration with the Department of Parks and Recreation, US Fish and Wildlife Service, and several departments at the University of California, Davis, they began developing and testing a barrier design that would protect the Shastas from the invaders just a few hundred feet downstream. Countless tests showed that the structure needed to be slick and without any seams for potential grip. During testing, Signal Crayfish were able to use any crack or crevice to gain a grip to climb up or around a barrier.

The final tests indicated that an appropriate barrier would be a one-foot-tall stainless-steel wall, with a four-inch overhang placed across the entire 150-foot width of the river. Tragically though, one year before the barrier was erected, the first Signal Crayfish were documented upstream of where the barrier was to be placed. Despite the private landowner's willingness to have this heavy-duty crayfish-proof barrier constructed at the edge of their backyard, the design, testing, and permitting took several years—enough time for the Signal Crayfish invasion front to reach this last isolated population of Shastas. Since then, there are no remaining *naturally occurring* populations of Shasta Crayfish that are unaffected by their invasive cousins.

Before, during, and after the Signals were documented and the barrier was constructed, snorkel surveys were conducted to remove Signal Crayfish.

The constructed barrier at one of the secluded springs that houses Shasta Crayfish (*Pacifastacus fortis*). Photo by Koen Breedveld.

But once the Signals get in, they are impossible to remove. The barrier keeps new Signals from invading and helps control the Signal population, but the best that can be done once Signals have entered an area is to keep their numbers as small as possible.

The decline of Shasta Crayfish in this area has been well documented. Slowly but surely, Signal Crayfish above the barrier began to displace the Shastas. Before the invasion, the lava boulders in the fish trap housed only Shasta Crayfish. A year after the invasion began, only Signal Crayfish were found in this isolated area. This replacement occurred on a small, well-documented timescale and in a small area. Repeat this process in every population of Shasta Crayfish, and these are the current circumstances this species is facing.

My time snorkeling in the secluded spring demonstrates the dire situation the Shasta Crayfish population faces. We collected only four Shasta Crayfish during the several-hour survey of this location, all within a small area immediately upstream of the barrier. The dense, deep gravel boulders in this area are the only section of the spring where Shastas are regularly found. Nearly every other section is overrun with Signals.

DISHEARTENED BY THE REPERCUSSIONS of delaying the barrier construction, Maria and Jeff knew that they had to get more creative. Now, there was not a single population of Shasta Crayfish living in safety from Signal Crayfish. And if Maria's and Jeff's years of observations at the fish trap were an indicator of what was to come, populations of Shastas would slowly wither away while Signals continued to multiply. If they can't remove Signal Crayfish from a Shasta habitat, Maria and Jeff figured that the next best thing would be to move Shastas to an isolated location that would serve as a refuge.

They found a location that would work—the best of three or four possible sites identified to date. This location had once held a thriving population of Shasta Crayfish, but the creek was diverted and completely dried up about seventy years ago.

Maria and Jeff figured that if the water that once flowed through this creek were returned to just part of its upper reach, this would create an area that could house an isolated population of Shasta Crayfish that was uninvadable, because downstream there was a steep drop-structure in a nearly dry channel—an impassible barrier to any crayfish. Without any Shasta habitats left uninvaded, the best option for the Shasta Crayfish is the creation (or restoration) of completely new refuge habitats. This potential refuge site could be a habitat for hundreds of Shasta Crayfish—a number that Maria isn't sure even exists in the dozen or so remaining naturally occurring scattered populations.

The restoration of this refuge area and relocation of Shasta Crayfish was a nearly twenty-year process. Maria and Jeff first came up with the idea when they lived in a small house only a few steps away from the once-flowing stream channel. To boil down twenty years of drama, permits, emails, headaches, and meetings into a sentence, they were able to restore the upper section of the creek to its former glory by moving the water diversion point downstream several hundred feet. With this restoration, this section of creek became one of just a few natural water bodies suitable for Shasta Crayfish but impossible for Signal Crayfish to gain access to. If Shastas could successfully be transplanted and settled into this new habitat, it could mark a new era in the conservation of the species.

Having been dewatered for more than half a century, the newly rewatered channel habitat needed to settle in and get back to its former condition before any Shastas were transplanted. Studies on the genetics of Shasta Crayfish populations demonstrated that one nearby population had the most diverse pool of genes, and the idea was to take Shastas from that resilient population (which had been invaded by Signals) and place them into this uninvaded refuge habitat.

In 2019, the stream segment was finally deemed ready for Shasta Crayfish—macrophytes were present, freshwater insects were seen scurrying between the rocks, and of course, the snails were also there. With the stream now an ideal Shasta Crayfish habitat, twenty-eight adult and juvenile Shastas were placed into the restoration site in 2019 and another forty-three in 2020.

The hopes are that this population will become a stable, reproducing population of Shasta Crayfish that can live their lives without interference from invasion. As of September 2022, a single sampling event has taken place at this location, and it seems that Maria's and Jeff's hopes are coming to life, as Shasta Crayfish of all age classes were detected at this site, including an ovigerous (i.e., with eggs) female. They found fewer Shastas than they'd hoped, but they also didn't search the entire habitat thoroughly because they wanted to minimize disturbance.

The success of this refuge site restoration is nonetheless a monumental feat in crayfish biology. The Shasta Crayfish is one of the most studied crayfish in the world, despite its populations being in the worst shape. The situation is so dire that the only option now seems to be creating or restoring uninvadable habitats while hoping to bring other minds and resources into play to devise methods of removing or, perhaps, sterilizing Signal Crayfish while leaving Shasta Crayfish unharmed and in place. These are the best glimmers of hope for Shasta Crayfish right now. Without more of these refuge sites, the documented decline of Shastas will likely continue.

SINCE STARTING HER WORK with Shastas in 1990, Maria Ellis has seen it all. She arrived to California after the species was newly listed as being state and federally endangered, ushering in hope for the future conservation of this species. After only a few years working with Shastas, Maria was hopeful, as she was still finding previously unknown populations. But since this hopeful era, nearly all the locations that were once abundant with Shasta Crayfish now harbor nothing but Signals.

The future of the Shasta Crayfish is perilous. Without a silver-bullet antidote that kills Signal Crayfish without harming Shasta Crayfish, the species-wide decline has no end in sight. Creation and restoration of refuge populations may be the species' only chance. Plans for captive breeding are currently underway, and a head start program would significantly boost numbers in isolated populations. Raising Shasta Crayfish until they are near the age of reproductive maturity in a laboratory, and then placing them into natural environments, would likely increase their reproductive potential by getting them past the earliest and most dangerous stage of a crayfish's life.

Although the genus *Pacifastacus* only contains six living species, the problems they face represent those that many other crayfish face. The five species that require conservation attention, *P. fortis*, *P. gambelii*, *P. connectens*, *P. malheurensis*, and *P. okanagaensis*, however, have at least one thing going for them: They are adapted to habitats that are often far from civilization and human activity (the same could not be said for *P. nigrescens*). The streams, creeks, and rivers these three imperiled species inhabit are well maintained, with minimal trash or pollution. The same cannot be said for the habitats of many of the imperiled crayfishes on the other side of the United States, many of which are just as likely to be found under a discarded car tire in a stream as they are under a rock.

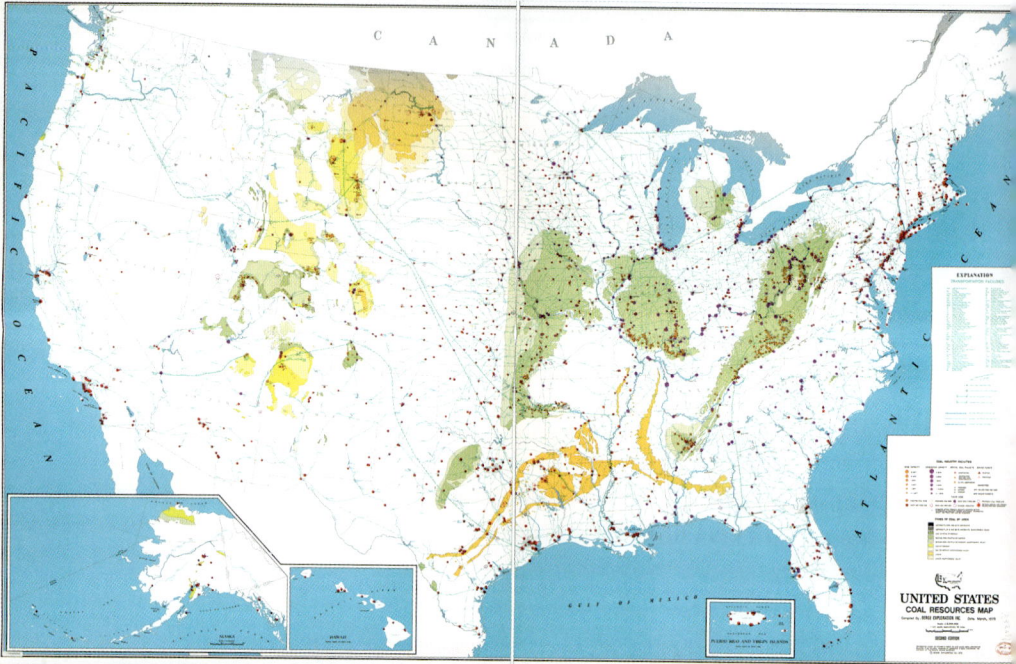

A map showing the coal resources (green and yellow areas) in the United States. The dominant shaded-green strip of coal resources throughout the Appalachian Mountains overlaps with some of the most diverse freshwater ecosystems in the country. City goers are often far removed from the reality of these industries, which take advantage of isolated, underdeveloped, but resource-rich areas throughout Appalachia. Wikimedia Commons.

# CHAPTER 8

*Cambarus callainus* and *Cambarus veteranus*

. . . . . . . . . .

## THE CRAYFISH
## IN THE COALFIELDS

WHEN WADING IN THE Tug Fork River, once you get knee-high you start to lose sight of your boots in the hazy water below. Even when the water is at its clearest, it's difficult to see through the rocks and refuse scattered all over the streambed. A five-minute wade in the water and a quick search of the stream bank would reveal enough tires, hubcaps, radiators, and mufflers to piece together a decent ride. These waters are always clouded and littered to some degree, but it can get much worse here and in the other streams in the coalfields located at the three-way junction of southern West Virginia, eastern Kentucky, and southwestern Virginia: After a major rain event, tiny particles called coal fines are pushed into the water, transforming it into a grayish-black haze.

Wolf Creek in Martin County, Kentucky, after an impoundment of coal slurry (a mixture of coal and liquids from coal preparation plants) broke into a mine in October 2000. This spill sent 306 million gallons of slurry down two tributaries of the Tug Fork River. Wikimedia Commons.

Coal fines are responsible for the debilitating respiratory disease known as black lung, which affects many people who have spent a lifetime working in coal mines. These small particles of coal do not easily dissipate in air or water, and when they are flushed into valleys, they blacken the water for days.

Only a few centuries ago, the waters in this area were tranquil and untainted. However, coalfield ecosystems are experiencing rapid transformations. Land-dwelling animals have the luxury of escaping polluted environments by simply running away, seeking better habitats. Deer, raccoons, opossums, and beavers can traverse great distances, even crossing mountains. Aquatic creatures like minnows, sculpins, hellgrammites, and crayfish are confined to heavily polluted waters, with no way to escape.

In many ways, crayfish are like little aquatic tanks. They have a hardened exoskeleton that protects their soft insides. And they are capable of surviving toe-to-toe battles with animals far outside their weight class, like frogs or snakes. During unexpected freezing temperatures, crayfish dig down in search of warmer, frost-resistant safety. Remember, crayfish have survived multiple mass extinction events. In adverse situations, crayfish have ways to prevail. This grit is the reason they have dominated freshwater habitats around the world.

But unfortunately, their resilience is being put to the test in coalfields, because the explosion of the mining industry (as well as of the tops of mountains, as you will see) has led to changes almost overnight to the streams, creeks, and rivers that crayfish have called home since long before humans were in the picture. Water chemistry in mine-adjacent streams can go haywire, and in some cases streams become completely smothered by rock overburden from the operations occurring above them. Even building infrastructure for mining can be disastrous for streams, because the construction of roads and bridges pushes sediments into the water, clogging up the cracks and crevices that crayfish love. With all these factors taken together, these environments are changing so fast that the crayfish are struggling to survive. For some species, extinction is a real possibility.

Saving crayfish from the megaindustry of mining and the associated threats of coalfields is no easy task. Luckily, wherever there are crayfish, there are dedicated and eccentric astacologists not too far away.

## Rediscovering the Guyandotte River Crayfish

Most states have a "crayfish person"—the academic, wildlife biologist, or even part-time hobbyist who holds encyclopedic knowledge on the distribution, biology, and conservation of their state's crayfish. Some states have multiple crayfish people, while others shift from one person to another. In his home state of West Virginia, Zac Loughman (pronounced "loaf-men") is currently the unofficial crayfish person.

Loughman is a tall, often heavily bearded biologist with a booming voice perfect for captivating an audience. Physically, he is far from quiet or gentle. You always hear him coming before you see him, whether it be his heavy footsteps or his loud exhalations. His interest in natural history started when he was an animal-obsessed child who accumulated animal books and went trudging through the woods anytime it was an option. Now, he can name pretty much any animal on a West Virginia hike—with scientific names and tidbits of natural history thrown in. He wasn't always into crayfish—an early passion of his was herpetology, the study of reptiles and amphibians. But while doing his master's research at Marshall University, he learned how difficult and competitive working with his favorite animals could be, so he switched his focus to crayfish.

I think it's safe to say that Loughman has flipped more rocks and dug more burrows than anyone else in West Virginia—an impressive feat, as several dedicated astacologists before him, including Raymond Jezerinac, Roger Thoma, Whitney Stocker, and even Curtis Newcombe back in 1929, spent significant portions of their lives working in the Mountain State. Many of the thirty-two native crayfish species in West Virginia have healthy widespread populations. But for others, their status has never looked worse. Loughman knows that one or two more serious disasters, like another ill-placed coal slurry spill or an unsuspecting bridge collapse, could wipe out an entire species.

After receiving his master's degree from Marshall, Loughman was given an opportunity to pursue his PhD at Indiana State University, while also teaching full time at West Liberty University, his alma mater in the northern panhandle of West Virginia. Having officially made the jump from reptiles and amphibians to crayfish, he needed a multiyear project worthy of becoming his PhD dissertation. His intuition to switch to crayfish was right, because he was able to receive funding to conduct a thorough, up-to-date, statewide survey of West Virginia's crayfish, together with his collaborator Stuart Welsh at West Virginia University. Surveys like this are a critical first step in conserving a state's wildlife, as they provide baseline information on how common or rare each species is, which species have been expanding their range, and which species are on the decline. Historic surveys in West Virginia are plentiful, which makes it easy to see how distributions have changed. Each time someone wanders the hills and hollers of West Virginia, they make new discoveries. But many of these modern discoveries are more doom and gloom than hoot and holler.

With his new teaching gig and access to eager West Liberty students, Loughman was able to access an army of undergraduate researchers who could aid with the statewide survey. Lucky for him, it's not that difficult to entice biology students to go play around in a stream. From 2007 to 2013, Loughman and his crew sampled a whopping 4,000 locations throughout the state.

At thousands of sites, Loughman and his students went from stream to stream and ditch to ditch flipping rocks and plunging burrows. Loughman eventually ended up in the southern portion of West Virginia, where he knew the stakes were much higher. Southern West Virginia is in the heart of the central range of the Appalachian Mountains, which means traversing only a few miles' worth of road may take half an hour due to the constant incline changes and hairpin turns. In this region, the only booming business comes from extractive industries, like mining and logging.

# WHAT'S IN A NAME?

When you conduct a multithousand-site survey, you often end up with some big discoveries, including new species. In West Virginia alone, Zac Loughman has described six crayfish species that were previously unrecognized. The taxonomic code (a strict and regimented set of rules designed to keep the science of taxonomy pure) states that describers are allowed to bestow a fitting name on a new species, which gets etched into the history books. It is common for names to honor the discoverer or an influential biologist in the field of study. In Loughman's case, some of the species he described are named in honor of crayfish biologists, such as the Allegheny Mountain Mudbug (*Cambarus fetzneri*), named for Jim Fetzner, Loughman's collaborator and the invertebrates curator at the Carnegie Museum of Natural History. Loughman named others after his mentors, such as the Meadow River Mudbug (*Cambarus pauleyi*), named for Tom Pauley, who was Loughman's master's adviser, who often let him chase crawdads instead of salamanders and snakes. Loughman even named some crayfish after iconic events in West Virginia history, such as the Tug Valley Crayfish (*Cambarus hatfieldi*), named for the historical Hatfield–McCoy feud in West Virginia and Kentucky. And eventually, some of Loughman's students even honored their leader with a species they described to thank and commemorate Loughman's dedication to West Virginia's crayfish: the Blue Teays Mudbug (*Cambarus loughmani*).

A Blue Teays Mudbug (*Cambarus loughmani*). Photo by Zackary A. Graham.

The Guyandotte River Crayfish (*Cambarus veteranus*) grows to the size of a hand, is chestnut to greenish-brown colored, and has orange-red bumps throughout, a spine on the side of its body, and big claws that you do not want to mess with. Photo by Zackary A. Graham.

In 2001, well before Loughman ventured down to the southern coalfield region, a separate group of biologists out of several West Virginia and Virginia universities conducted a survey for what was at the time thought to be the rarest crayfish in West Virginia, the Guyandotte River Crayfish (*Cambarus veteranus*), named for the 166-mile-long river that flows through the heart of the coalfields. At the time, *C. veteranus* was thought to reside in West Virginia, Kentucky, and Virginia. In West Virginia, *C. veteranus* was known in only three counties. It had always been rare, but no one realized just how rare until someone went out to look.

The 2001 survey searched fifteen West Virginia locations where this species historically lived. Not a single *C. veteranus* was found, which led the authors to conclude that things were not looking good for this species in West Virginia. Streams once filled with prime habitat were now desolate and buried under a quarter-inch layer of sediment. Since the last survey, the infrastructure for extractive industries had started wrecking the streams where *C. veteranus* once thrived. The resulting paper concluded that *C. veteranus* had been wiped out from West Virginia, with the last known collection occurring in 1989—a twelve-year disappearance.

What causes a species to decline in such a short timescale? Simply put, the streams where *C. veteranus* had once thrived were becoming overrun and clogged with sediment. No single source of sediment can be given the blame, but rather there are many culprits. The two biggest responsible parties are mining and logging operations, which require big machinery that mountain roads are not used to handling. These roads are small and need constant repair. When a mining operation swings into action and the infrastructure changes begin, trouble flows downstream.

The streams become inundated with an onslaught of minuscule sediment particles. It's like a torrential downpour of muck and silt, which suffocates the very places where our crayfish friends love to burrow and seek refuge. Every nook and cranny that crayfish rely on becomes a muddy grave, snuffing out their chances of survival. The streams that were once vibrant and full of life now face an uncertain future, as sediment suffocates their delicate ecosystems. The comorbidities caused by these changes ripple through the animal kingdom, leaving a trail of struggle and hardship in their wake. Humans must understand the impact of our actions and strive to find a balance, for the survival of species like *C. veteranus* depends on it.

IN 2009, WHEN IT WAS Loughman's time to drive south and search for *C. veteranus*, the species had not been seen for twenty-years in West Virginia. He, like others, assumed that this species had disappeared in the state. But he knew that if he could find just one or two populations in middle-of-nowhere streams, there was still hope for this crayfish. So with the goal of rediscovering *C. veteranus* and making waves in the crayfish community, Loughman and his crew loaded up in a fifteen-passenger van to start their quest.

With a daunting task at hand, Loughman and his students drove from stream to stream in search of *C. veteranus*. The goal was to search anywhere and everywhere throughout the two river basins historically known to house *C. veteranus*—from heavily trafficked populated streams to remote areas that take ten miles to hike into and have likely never had their rocks flipped.

As the van doors flung open at a new site, the students quickly gathered nets and bounded into the streams, while Loughman observed from the stream bank and filled out data sheets about the location, collecting GPS coordinates, habitat quality data, and notes on human disturbance. Finding crayfish in these streams isn't hard, because some species in the region are dirt common, like the Spiny Stream Crayfish (*Faxonius cristavarius*) and the Coalfields Crayfish (*Cambarus theepiensis*), another species described by Loughman. But finding *C. veteranus* was proving to be a bigger task than Loughman had expected.

After nearly 100 failed stream searches, the typical motivation tactics started to lose their appeal and student morale was tanking. So Loughman increased the stakes: He would let his students cut his hair into a mohawk if they were to find a *C. veteranus*. Sporting a mohawk is far from his natural look—so the hilarity of the situation motivated his students to keep trucking from site to site.

Anxiety peaked as Loughman and his students pulled up to Pinnacle Creek, a site that will go down in history books; twenty years earlier, this site was one of the last known locations in which a West Virginia *C. veteranus* was found. Pulling up on this site, Loughman was on high alert because he knew this creek could be one of his last chances to find the lost species. Pinnacle Creek meanders through the mountains and flows all the way downstream into the town of Pineville. In the Pineville section of Pinnacle Creek, there is clear evidence of humans everywhere you look: tarps, car parts, invasive species galore, you name it.

But the upstream section of Pinnacle Creek looked like crayfish paradise. Loughman is a naturalist at heart, and skilled naturalists can do things like listen to bird calls, associate them with the species of trees in the area, and then interpolate how the tree community can impact the hydrology and flow of the stream below—an almost wizard-like skill set that can be acquired only after dedicating years of one's life to nature.

From looks alone, Loughman knew that this habitat was ideal for *C. veteranus*: The water flowed with clear boundaries between the white-water riffles, clear runs, and deep pools. It is well-known that *C. veteranus*'s favorite habitat is among medium to large rocks in rapidly moving water, which is always where Loughman directs the effort when approaching a site.

Yet, again, after clearing out the rocks in the crayfish's favored riffles and runs, there was still no sign of *C. veteranus*. Before packing up and heading back to the bumpy roads, Loughman directed his students toward one final rock. It was long, skinny, and awkward—about three or four feet long, appearing like a petrified railroad tie embedded in the stream bank. This was far from the preferred habitat of *C. veteranus*, but Loughman knew that big rocks equal big crayfish, so it was worth a try. After some prying and the standard swooshing of water, the crew picked up the net and saw a career-defining crayfish.

Loughman instantly recognized the crayfish in the net, though 99.99% of people on earth would not. Orange-red highlights, large greenish-brown claws, and spines on the side of its body made this unmistakably a *C. veteranus*, the first time one had been collected in Loughman's home state in over twenty years. In this moment, Loughman knew this was going to be the start of a lifelong journey. Because, now that he knew *C. veteranus* was indeed still alive (but clearly not thriving) in these remote reaches of Pinnacle Creek, it suggested the species could be in other seldom checked areas as well. Loughman's goal was now to protect this species with federal jurisdiction—the only way to protect this animal and provide the support it needs to recover.

After inspecting the crayfish from the rostrum to the telson, Loughman finally came back to reality only to realize his students were laughing uncontrollably. While Loughman had been realizing that his work with this species had only begun, his students had started planning the mohawk they now got to shave onto their professor's head.

## Saving the Crayfish in the Coalfields

Even though rediscovering *C. veteranus* in West Virginia was career-defining for Loughman, it wasn't long until one newly discovered population of *C. veteranus* turned into two—and then two populations turned into three. Throughout the surveys, Loughman and his crew found nearly a dozen West Virginia records for this species. Even though he was ecstatic that *C. veteranus* was still alive, he knew this animal was declining.

The goalposts quickly shifted from proving that *C. veteranus* was still around to getting this animal the protection it needs to stay around. For this to happen, *C. veteranus* needed to receive federal protection under the Endangered Species Act (ESA). When a species is listed under the ESA, it is provided with the support it needs to be properly managed and conserved. When my parents were children, bald eagles were nearly extinct, but now they are thriving and have stable populations throughout the United States—all thanks to the ESA. Humpback whales, Florida manatees, California condors, and many more have all survived because of the ESA and the protection that it brings.

Once a species gets listed under the ESA, the US Fish and Wildlife Service (USFWS) has a duty to do whatever it can to protect that organism. But most of the success stories or poster children of the ESA are cute and cuddly animals. Getting a beady-eyed, ten-legged crayfish listed under the ESA is not easy. In the early 2010s when all of this *C. veteranus* work was happening, only four species of crayfish had ever been listed under the ESA (one of them being chapter 7's Shasta Crayfish). Four endangered crayfish may seem like a lot, but it is laughable compared to the numbers of other imperiled freshwater animals; freshwater mussels have nearly eighty species protected by the ESA.

Loughman believed that *C. veteranus* was a shoo-in with the ESA; *C. veteranus* is an invertebrate poster child for the ecological nightmares that occur in coalfields. And protecting this animal would provide an umbrella of protection for the other little critters that call this area home. Coal mining is a top threat to the animals in this secluded section of Appalachia, with coal mines surrounding all known populations of *C. veteranus*.

COAL IS TRADITIONALLY MINED using two different methods: deep mining and surface mining. In the early days of mining, deep mining was prominent. This is the classical depiction of miners walking into a damp, dark mineshaft and extracting minerals from a mountain's core. Although exceedingly harmful and dangerous to the humans who work in the mines, deep mining causes relatively minimal change to the surface landscape. But with surface mining, alterations to the environment are major. Several techniques fall under the umbrella of surface mining that you may have heard of: strip mining, open-pit mining, highwall mining, and of course, mountaintop removal. Surface methods are preferred in the modern era of mining because they are more cost effective.

Although cost effective, surface mining is exorbitantly inefficient. This is because coal is not found in massive, easily harvestable chunks but in thin horizontal layers embedded in mountains, called coal seams. In many cases, to access a single coal seam, you have to remove roughly 99 meters (324 ft) of rock. A football field must be carved out of the mountain just to access a small strip of coal. And because seams are relatively scattered, once they are depleted, the operation packs up shop and moves to another mountain. This is exactly what has happened throughout Central Appalachia over the last half century. Surface mining operations set up, extract what they can until the resources are depleted, and move on to the next location. Ten percent of the land in Central Appalachia is either actively being surface mined or is reclaimed surface mining land. A half-hour drive from the university I work at in the northern panhandle of West Virginia can land me in the middle of at least half a dozen surface mines.

Many of the environmental issues from surface mining stem from this practice's inefficiency. Surface mines can be expansive—hundreds of square yards in size. All the excess non-coal rock material that is either exploded or dug out, called overburden, needs to go somewhere. Overburden is typically dumped into large pits, which are spread back throughout the mined area.

# THE DECAPITATION
# OF MOUNTAINS

If you find yourself wandering through the central portion of the Appalachian Mountains, you might stumble upon a practice that could shake you to your core. They call it "mountaintop-removal mining," which sounds like a technical and polite term, but once you witness it firsthand, you realize how extreme it is. One minute you might be following a picturesque mountain stream, and the next minute you find yourself in a barren, gray landscape filled with heavy machinery. This practice involves removing the tops of mountains to reveal the precious minerals and coal beneath. Forests are clear-cut and burned, and explosives are placed to blast away the mountaintops.

The issue with this practice is that it's not just a matter of removing the mountaintops; it's more like a decapitation. Mountains take tens of millions of years to form, and the mountaintops throughout Appalachia are ancient, formed over 270 million years ago. These mountaintops are being blasted away, and they're not going to regrow anytime soon.

The aftermath of mountaintop removal. Wikimedia Commons.

But a lot of this overburden starts to creep down the mountainside. Just like water, overburden follows the laws of gravity and will eventually make it down to the lowest-elevation areas: streams, creeks, and rivers.

When overburden makes its way down the mountain, it ends up clogging waterways below, removing any small cracks and crevices that are necessary for the survival of the animals that need these spaces for refuge. And because the overburden comes from mountaintops, there is a massive input of foreign chemicals from these rocks into streambeds, altering water chemistry beyond their natural conditions. Over 2,000 miles of Appalachian streams have been given an overburden burial.

The sedimentation brought about by mining is a major threat to *C. veteranus* and the other wildlife in this region, but it isn't the sole contributor to the animals' downfall. Any practice that involves major construction events (i.e., using monstrous machines near remote mountain streams) will clog up the stream below and alter its water chemistry. Logging and highway construction are two other obvious infrastructure culprits. On top of this, all-terrain vehicles (ATVs) are probably responsible for squishing a few rare crayfish ever year. ATV tourism is one of the only economic stimuli in much of this region. And with ATVers, many trail riders' favorite thing is to fly right through a stream, which risks potentially crushing West Virginia's rarest crayfish, not to mention disturbing sediments by fording the stream.

With a disastrous combination of interrelated threats—coal mining, sedimentation, logging, physiochemical changes in streams, ATVs, and highway construction—everyone assumed that *C. veteranus* would get protection from the ESA, but nothing in biology is ever so straightforward.

AFTER MAKING THE BIGGEST FIND of his career, Loughman was surprised to find out that, at the end of the day, conservation action was not coming as soon as he had hoped. At this stage, crayfish conservation was in its infancy in West Virginia, and agencies like the USFWS had not put crayfish high on their radar. For years after his rediscovery, Loughman went parading around from one conservation meeting to the next presenting on *C. veteranus*, but no one seemed to care. And there was no movement from ESA officials regarding the status of *C. veteranus*.

But all of that changed in 2010, when people started paying attention to a campaign known as the Southeast Freshwater Extinction Crisis, powered by the Center for Biological Diversity, a nonprofit conservation organization. This campaign, designed to protect the Southeast's "unparalleled aquatic biodiversity," came with a list of 404 aquatic species that deserved threatened or endangered status under the ESA. And of the 404 species highlighted, there were a whopping eighty-one crayfishes from the thirteen southeasternmost states. This document recognized crayfish as one of the Southeast's prized animal groups—on the same level as other historically imperiled groups like freshwater mussels, salamanders, and darters.

Cambarus veteranus was one of those eighty-one crayfishes, and this proposal (plus the subsequent legal battle between the Center for Biological Diversity and the USFWS) eventually brought C. veteranus before the eyes of the USFWS. For a species to become listed under the ESA, it is slightly more complicated than filling out a set of forms and waiting in line at some government building. The process is grueling, as the USFWS gathers and evaluates all the evidence on a species' conservation status and risks of imperilment. Periods of expert review are followed by periods of public comment— federal taxpayer dollars are used to protect species listed under the ESA.

Before any decisions are made, the USFWS needs all the information it can get on a species, which is done through a species status assessment, or SSA. SSAs contain everything you need to know about a species: where they live, how many individuals are in each location, how often they reproduce, what their threats are, and so on. SSAs are typically conducted by USFWS employees. But in 2014, the USFWS had no crayfish experts on board, so Loughman was tasked with assisting in the creation of the SSA for C. veteranus.

Even with a path to federal listing in place, thanks to the Center for Biological Diversity and to the USFWS's willingness to protect this crayfish, there was one big issue that needed to be addressed—one which Loughman and his colleagues feared would prevent C. veteranus's listing and protection. Cambarus veteranus historically resided in both the Upper Guyandotte River Basin and the Big Sandy River Basin. Out of all the populations of C. veteranus that Loughman found in West Virginia, only one population was from the Upper Guyandotte River Basin. The rest were from the Big Sandy River Basin. And when Loughman and his crayfish mentor Roger Thoma were examining the Big Sandy River Basin crayfish, they looked . . . different. Different enough to get that lingering thought in their heads as to whether they were even the same species.

This issue created taxonomic and governmental mayhem, because C. veteranus was in the process of potentially becoming federally endangered. But there was a suspicion that these two populations were in fact two separate species masquerading as one. Splitting C. veteranus into two species would mean that the entire range of the species would also be split, leaving not just one species with serious conservation concerns but two. If that happened, the crayfish from the Upper Guyandotte River Basin would be extremely rare, whereas the Big Sandy River Basin crayfish would have a dozen or so known populations.

To wrap up a long and convoluted story, which deserves a book in itself, Thoma, Loughman, and molecular geneticist Jim Fetzner believed that there was enough evidence to split the two disjunct populations previously known as *Cambarus veteranus* into two separate species. The populations from the Upper Guyandotte River Basin remained *C. veteranus*. At the time, there was only one known location where this species occurred: Pinnacle Creek. The populations formerly recognized as *C. veteranus* from the Big Sandy River Basin were described as a new species, the Big Sandy Crayfish (*Cambarus callainus*). And just like that, West Virginia had its newest crayfish. Now the conservation issues of one species had essentially doubled—with both species requiring protection if they were to survive the ecological hardships of the coalfields.

During the public comment period for the SSA, the local mining and logging industries highlighted three major criticisms. First, they didn't believe that *C. veteranus* and *C. callainus* were indeed separate species, and that it was just taxonomic wizardry hoping to hamper the industries' spread in Appalachia. Second, they argued that neither species was as rare as the SSA had indicated. And third, they suggested that there wasn't enough evidence for what was causing the species' imperilment.

These critiques brought on what Loughman affectionately refers to as the "summer of hell," because Loughman and his team needed to respond to these comments in order to put forth the arguments for their federal listing. So over a three-month period in 2015, Loughman and his students had to traverse the coalfields and take samples throughout the entire known range of *C. veteranus* and then come back home to process data and measure crayfish immediately—only to repeat the entire process the next week. These efforts found *C. veteranus* in the Clear Fork watershed of the Guyandotte River, increasing the species distribution to two named streams in West Virginia. Twelve-hour days in the field were immediately followed by twelve-hour days in the lab. And this "grind and slog," as Loughman described it, continued for weeks, until eighty or so locations had been thoroughly sampled for *C. veteranus*. The faster this data was collected, the sooner *C. veteranus* and *C. callainus* could be protected. In the end, seventeen streams were determined to house *C. callainus* in southwestern West Virginia, Eastern Kentucky, and southwestern Virginia.

Loughman eventually collected the data he needed to respond to the public comments and turned the data over to the USFWS. He showed, with current, updated analyses, that *C. veteranus* and *C. callainus* were in fact sep-

# BIG SANDY CRAYFISH
## (*CAMBARUS CALLAINUS*)

Compared to its sister taxon, *Cambarus veteranus*, *C. callainus* grows to larger sizes and can often be identified by its turquoise-green tint—especially after a fresh molt, when it may appear to be highlighter blue. The claws and rostrum also differ between the two species, with typical *C. callainus* claws being longer and thinner, showing potential adaptations to live in the faster-moving streams in the Big Sandy River Basin. The species name *C. callainus* comes from the Latin word for turquoise green, in reference to the claw color.

A Big Sandy Crayfish (*Cambarus callainus*).

arate species, that they were rare, and that throughout their range, habitat degradation and increases in sedimentation are likely the main driver of their decline.

IN APRIL 2016, after a five-year crayfish-conservation roller coaster, the USFWS released a thirty-three-page document pronouncing its ruling on the two coalfield crayfishes. The federal government declared that under the ESA, *C. callainus* would be listed as federally threatened, whereas its two-stream cousin *C. veteranus* would be listed as federally endangered. Endangered status receives the highest level of protection, as such species are deemed at serious risk of becoming extinct without intervention. One step down from endangered is threatened status, which is for species that are likely to become endangered in the foreseeable future.

Not coming as a surprise to anyone who has sampled these species, the USFWS also determined that increase in sedimentation and alteration of streams' natural water quality were the top threats to these crayfish. It now became the government's obligation to conserve these species, the fifth and sixth crayfish species ever protected by the ESA. Preserving these species takes many forms, including monitoring and surveying the health of their populations as well as supporting efforts to minimize environmental impact in their area. Gaining ESA protection meant that in October 2016, USFWS officials launched plans to block the dumping of mining waste into watersheds sustaining the endangered *C. veteranus*, whose range had shrunk by 92%, and the threatened *C. callainus*, which had experienced a similarly large decline of 62%.

Without further disturbance, *C. callainus* may be able to make a quick exit from the ESA. *Cambarus callainus* has a handful of populations in southern West Virginia, eastern Kentucky, and southwestern Virginia, some of which are doing better than others. *Camabarus veteranus*, on the other hand, is in dire straits, with only two small populations, both of which have bleak numbers. *Cambarus veteranus* is in such bad shape that all research is currently on pause because both populations are so fragile.

Luckily, both *C. veteranus* and *C. callainus* now have support from dozens of groups, including the USFWS, West Virginia Department of Natural Resources, West Virginia Department of Highways, and of course, Loughman and his student army at West Liberty University. These types of collaborative efforts are crucial, because the struggles that these crayfish face are not one-dimensional. High above the coalfield streams remains the imminent threat of mining. But one of the biggest and more recent threats occurs right above the streams, from a "necessary evil" of infrastructure: bridges.

# Bridges and the Big Sandy Crayfish

When you see nearly twenty adults spread out in a stream wearing aquatic attire and wielding Little League–sized soccer nets in thigh-deep water, you know something is up. Even in a remote West Virginia stream, this spectacle attracts attention from locals wondering what all the hubbub is about. From an onlooker's perspective, it may seem like a nonchalant, fun event, where a bunch of adults get to play in the water. But in the water, everyone's movements appear programmed and intentional, which mirrors the tense and anxious feeling in the air.

The mission of this controlled chaos was to collect and move as many *Cambarus callainus* as possible. Within a few weeks of this crisp July morning, the Big Sandy Truss, a roughly sixty-foot-long, two-lane bridge that spans the Tug Fork River, was going to undergo structural improvements. To make matters more complex, under the Big Sandy Truss lived one of the strongest populations of *C. callainus*. A massive collaborative effort between private consulting companies, public universities, and both state and federal government agencies aimed to prevent the unnecessary death of the federally threatened *C. callainus*.

Loughman and his students at West Liberty University formed nearly two-thirds of the workforce at this event, which was the second of several events planned around secluded West Virginia bridge reconstruction projects. Loughman serves as the species expert for *C. veteranus* and *C. callainus* and acts as a major force in coordinating what he calls "boots on the ground" conservation. Without people physically flipping every rock, collecting, and then relocating every single *C. callainus* beneath the bridge, their chances of survival would be minuscule.

Bridges are an essential part of modern life and a testament to human engineering. However, to crayfish, bridges spell trouble. Underneath every bridge there was once a thriving, well-balanced aquatic ecosystem. But when a bridge is constructed or renovated, the balance becomes skewed. Bulldozers, excavators, and cranes are brought in to clear the surrounding area and install the necessary infrastructure; this machinery often sits above or on the sides of stream banks—causing sediments from construction to fall straight into the water below. These sediments create thousands of little landslides that clog the water, making it difficult for animals like crayfish to burrow.

While precautions can be taken to mitigate the necessary evil of bridges—such as silt curtains, which reduce sedimentation—they can only do so much. Disturbing the water below is unavoidable during a construction event of this scale. And throughout the ranges of (the now federally protected)

C. *callainus* and C. *veteranus*, many bridges are not up to modern standards. Many of these bridges requires structural updates that will undoubtedly cause disturbances to the crayfish below. For the local economy, with major coal mining, logging, and natural gas drilling occurring throughout the region, rebuilding these bridges is imperative.

LOOKING UPSTREAM from the old and rusted frame of the Big Sandy Truss, you would see plenty of pristine C. *callainus* habitat, including several riffles and runs peppered with slab boulders that, when flipped, often reveal a hefty C. *callainus*. Three months earlier, the first mass translocation event of a North American crayfish species allowed researchers to work out the kinks in conducting a daylong translocation event of this scale. But now, in July, the stakes were much higher: Rising temperatures in mid-to-late summer trigger the onset of C. *callainus*'s reproductive season. Now, collecting and saving the C. *callainus* could make or break the future of this stream's population.

Cambarus *callainus*, like many larger-bodied members of its genus, has a slow and steady reproductive cycle, reproducing only once per year. Growth data suggests that C. *callainus* live five years or more, and females produce just over 100 eggs each time. Contrast this with the live-fast-die-young life cycle of other crayfish species, such as the Red Swamp Crayfish (Procambarus *clarkii*), who may only live one or two years but produce five times the number of eggs per season compared to a C. *callainus*. Crayfish that produce fewer eggs less often, like C. *callainus*, are the ones that are most at risk of imperilment. With C. *callainus*'s breeding season in full swing, the July translocation at Big Sandy Truss would protect not only adults but also potentially thousands of little C. *callainus* craylings. Every step taken in the stream warranted extra caution: egg-carrying, or soon to be egg-carrying, females were in the forecast.

Planning this translocation for July was no mistake though, as this event was intended to kick-start a new era in North American crayfish conservation: The USFWS, ESA's managing institution, gave permission to collect up to twenty-four fertile C. *callainus* to send to White Sulphur Springs National Hatchery for captive holding. This hatchery, located in White Sulphur Springs, West Virginia, is famous for propagating rare aquatic wildlife, including highly threatened species like candy darters. Now it was going to get its first shot at caring for and propagating one of the rarest crayfish in the world.

On top of the twenty-four fertile females going to White Sulphur, an additional forty crayfish were to be taken to the facilities at West Liberty University for temporary housing while the Big Sandy Truss was being reconstructed. The number of crayfish to be taken was determined by the USFWS and calculated based on the best available data known for *C. callainus*. All other *C. callainus* collected—aside from those to be transported to White Sulphur and West Liberty—would be transported to safety a half mile upstream.

THE TRANSLOCATION kicked off right after sunrise, with the first rocks getting flipped at roughly eight o'clock. An early start was imperative, because with 200 meters (646 ft) of habitat to sample, and nearly every rock needing to be flipped, everyone was expecting to wake up with sore fingers and backs the following morning. The goal was to start immediately upstream of the bridge and collect every single *C. callainus* within that 200-meter stretch. Collecting reproductively active or egg-bearing females was the ultimate hope, but no one knew for sure that the timing was perfect until they started flipping rocks.

At sunrise, the cold chill in the air kept the mood somber. And the first hour or so of collecting did not yield hopeful results: The group had collected fewer than ten *C. callainus*—none of which were reproductive females. But everything started to change once the collectors reached about 50 meters (164 ft) upstream from the bridge, which marked the location of a riffle with tons of habitat to sample. In this area, nearly every flip revealed an adult *C. callainus*—apparent from the exclamations of joy coming from the crews every time they searched through their seine net.

A bystander may not have appreciated the effort and organization that was occurring. But in the water, everyone had their own role. Many swapped throughout the day between rock-flipping, some form of data collection, and a supervision role. In the stream itself were three separate groups of four individuals, each of which were taking turns flipping rocks, holding the seine nets, and checking for *C. callainus*, all while trying to keep their balance in the rushing turbulent water. On the stream bank, white foldable tables were set up like an aid station for marathon runners, but instead of cups of water, these tables had an assortment of tubs, bags, calipers, notebooks, and other supplies for formal data collection. Several people were acting as taxis, picking up crayfish from the collectors, recording data, and transporting the precious cargo to the data tables on the stream bank. Senior members of the group were on the bank, barking out orders and engaging in streamside politicking.

At the time, I was a postdoctoral researcher in Loughman's crayfish lab at West Liberty. Postdoctoral positions are now common middle-ground positions after one completes a PhD but needs more experience to become a professor. I was fortunate enough to get a position with Loughman straight after graduating from Arizona State. The role was perfect for me, because I was only an hour or so away from my hometown and also learning from one of the leading North American crayfish experts. It was days like this that I was able to demonstrate my abilities to Loughman and others, which thankfully landed me a professor position at West Liberty a year later.

On this translocation day, I was tasked with being one of the crayfish taxis—running bait buckets filled with crayfish from the stream to the data collection tables. In addition to taxiing, I collected GPS data from the exact location each crayfish was located. To do this, I was graced with a funny-looking hat hooked up to a shoulder-strap GPS that synced with my phone as the interface.

After six or so hours of sampling, with about 100 meters (328 ft) of stream having been searched, it became apparent that we had already collected the allotted number of females to be transported to West Liberty University and White Sulphur Springs. It actually turned out to be a best-case scenario for the females too, because none of the C. callainus had extruded their eggs yet, but they were all exhibiting active glair glands, which meant that they would be laying their eggs in the coming weeks. Egg extrusion is a stressful time, and transporting these crayfish with eggs would have been a nerve-racking experience. But because only their glair glands were active, they would have a stress-free transport to their temporary homes, where they would hope-fully lay eggs in captivity. Once the allotted number of glaired females were collected, two vehicles packed with coolers and one-gallon baggies, each containing a single crayfish, were driven directly to their respective facilities.

Once those females were on their way to spend the foreseeable future in human care, any additional crayfish collected just became a bonus, and we would shuttle them upstream away from the bridge and place them in another section of ideal habitat. As the day waned, the vibe switched from a stressful aura of military-style orders shouted from the stream bank to an aura of calm and proud energy as the C. callainus started rolling in. That day, after fourteen hours, we had collected a total of 185 C. callainus.

THESE TYPES OF TRANSLOCATION events were the first of their kind for any North American crayfish species, ushering in a new era of crayfish conservation. Radical and active conservation efforts like translocations are likely going to be a mainstay in the future of crayfish biology. In the past, protecting and maintaining habitat has always been seen as one of the best ways to preserve species. But some species are declining so quickly, with their environments changing faster than ever, that the old strategies of passive protection no longer work, and proactive "boots on the ground" conservation is the only way to protect our wildlife and the ecosystems in which they reside.

Not only were nearly 200 *C. callainus* saved that day, but the ones transported from the polluted and sedimented waters of the Big Sandy River to aquaculture facilities proved to be another monumental success. The female *C. callainus* taken to the White Sulphur Springs hatchery and to West Liberty University ended up extruding and laying eggs while under human care. The females raised those babies in captivity, so when the Big Sandy Truss was finally reconstructed, more crayfish were put back then had been taken.

Since 2021, four Big Sandy Crayfish translocations have occurred. The first was on April 16, 2021, underneath a railroad bridge, and 118 *C. callainus* were collected and moved upstream away from the repair work. The second event (July 23, 2021) was the one in which I partook, described above, with 185 *C. callainus* being either translocated upstream or taken into captivity for captive breeding. The third event (September 10, 2021) occurred underneath a much smaller bridge, called the Panther Girder; this habitat is not as pro-

A Big Sandy Crayfish (*Cambarus callainus*). Photo © Joel Sartore / Photo Ark.

ductive as the Big Sandy River, and thirty-two *C. callainus* were collected and
moved to safety that day. The fourth translocation (June 23, 2022) was also
underneath the Big Sandy Truss, with a whopping 255 *C. callainus* collected!

WORK WITH C. CALLAINUS serves as a textbook example for how the ESA
can protect a species. Without protection from the ESA, none of this work
with *C. callainus* would have been possible. And the successes of the captive-
reared *C. callainus*, both at White Sulphur Springs hatchery and at West Liberty University, are foreshadowing the future of crayfish conservation in
North America. Instead of sitting around and hoping that populations recover, we must initiate proactive recovery methods and breeding programs.
Similar breeding programs have been the saving grace of many vertebrate
species, but our understanding of crayfish rearing is in its infancy.

It is Loughman's hope to make *C. callainus* the first crayfish *removed* from
the ESA. This will require further protection from anything that increases
sedimentation, like mining, logging, or bridge reconstruction, but it will
also take furthering our understanding of captive rearing of crayfish. This
work with *C. callainus* may end up also helping its two-stream cousin *C. veteranus*, whose situation is dire, with work on this species coming to a halt
because of its low population sizes. It is unlikely that *C. veteranus* will ever be
able to recover without substantial human intervention.

Some species will continue to decline, whereas others will succeed.
The crayfish in the coalfields present a dichotomy that can shed light on
crayfish conservation issues throughout North America. Some species, like
*C. veteranus*, are in so much trouble that biologists fear that even stepping
in the stream would cause minor harm, whereas others, like *C. callainus*,
demonstrate what can be done to help them with proactive conservation.
The future of crayfish conservation is likely to throw additional curveballs
and certainly invite some new species onto the ESA. But one thing is clear:
Whether a species is being diminished by its invasive cousins or its streams
are being polluted to hell and back, having a "crayfish person" on your side
can be a species saver.

# CHAPTER 9

# Describing (and Conserving) the Future of Crayfish

## Hidden Diversity

WITH HIS HANDS buried deep in a burrow somewhere off the Gulf Coast, Mael Glon, a PhD student at the time, was expecting to plunge out a familiar mud-ridden Painted Devil Crayfish (*Lacunicambarus ludovicianus*)—a typically camo-green crayfish with orange highlights and a characteristic three stripes running down its abdomen. This was a species Glon had collected many times before, and he knew it was historically known in this area. But with some luck, he pulled out a crayfish with a brilliant turquoise hue, highlighted in fiery orange at every joint, and not a trace of the three stripes he was expecting. This was a creature so

The brilliant colors of the burrowing crayfish that Mael Glon dug up that day. Photo by Guenter Schuster.

colorful that it would be more at home in a child's coloring book than in the depths of a burrow. And it wasn't a fluke either, because every burrow from this location yielded a similar turquoise-orange beauty. As Glon gazed down at these crayfish, he felt a rush of confusion and excitement only familiar to the dedicated few that spend their lives thinking about a single group of organisms. He had just stumbled upon a mystery that would keep him occupied for years to come.

Glon was on a mission to unravel the taxonomic status of crayfishes in the genus *Lacunicambarus* during his PhD research at Ohio State University. He had spent countless hours traveling across the United States, from the Gulf Coast to the Great Lakes, digging up crayfishes and gathering specimens. And on this trip to the southern reaches of Mississippi and Alabama, Glon was hoping to add the final pieces to the puzzle by collecting specimens of the Painted Devil Crayfish and the Rusty Gravedigger (*Lacunicambarus miltus*). But finding this turquoise-and-orange-banded crayfish threw a wrench into the gears of Glon's dissertation, because a crayfish like this had never been recorded in the area.

In the days that followed, Glon and his colleagues continued their search for this newfound candy-colored crayfish. Glon had a hunch that this crayfish represented an undescribed species and was not just a weird variation of a Painted Devil Crayfish. They were just too different. Happening on undescribed species is more common than many expect, especially for groups

The only two known *Lacunicambarus* species from the Gulf Coast at the time were the Painted Devil Crayfish (L. *ludovicianus*; *top*) and the Rusty Gravedigger (L. *miltus*; *bottom*). Photos by Guenter Schuster.

that receive little attention, like burrowing crayfish. Thus, Glon's trip down south took a U-turn and now included a new goal: to gather more evidence to test the theory that this turquoise-and-orange crayfish was a new species. Despite their best efforts, the search proved challenging, and Glon struggled to map out the potential distribution of this mysterious creature. In the end, he found it at only nine locations, mostly near the Gulf of Mexico and at low elevations. And catching it was no easy feat, as most sites held only a few burrows and only one individual could be collected at a time. The real answers would not be revealed until genetic data was analyzed.

With only nine known locations, Glon knew that looking at these crayfishes' DNA would be critical. This data would determine whether the populations were different enough to qualify as a new species or if they were highly variable populations of the Painted Devil Crayfish. Taxonomists like Glon use morphological and sometimes geographic evidence to support their claims that one species is different from another. More recently, a new era of taxonomy has been ushered in and genetic data is more or less a requirement to support the claims of a new species.

Unsurprising to Glon, the DNA showed that all of these turquoise-and-orange-banded crayfish were closely related to one another but distantly related to other crayfish in the genus *Lacunicambarus*. Now knowing that this species was genetically distinct, Glon could combine this information with the data he'd gathered about their distribution as well as morphological data in an official species description.

Taxonomy and taxonomic descriptions are a labyrinth of language, where species descriptions are written in a code of intricate anatomical terms, such as "Carpus with prominent dorsal furrow and one weak dorsomesial tubercle" and "Proximal podomere of uropod with distal spine on mesial lobe." But when a biologist suggests that an animal is deserving of an entirely separate name, a high level of detail is required, which enables future biologists to be able to discern and differentiate between these animals. And this process, like all good science, undergoes peer review from other taxonomists, who review the evidence and ensure that the data indeed leads to the suggestion that a new species description is warranted.

In the final step of a species description, the author provides a common and scientific name for the species. Glon chose a fitting name for this new species: the Banded Mudbug. And the scientific name is *Lacunicambarus freudensteini*, attributed to one of Glon's PhD advisers, Dr. John Freudenstein. In his paper, Glon also described a second species that he focused on during his Gulf Coast collection trip, known as the Lonesome Gravedigger (*Lacunicambarus mobilensis*).

# SPECIES COMPLEXES

Popular media often depicts biologists journeying into the undiscovered parts of the Amazon or ocean depths to describe new, never-before-seen forms of life. Few realize that species new to science are everywhere. You may have interacted with undescribed species and not even known it. "Finding" a brand-new species is not common, but more realistically there are undescribed species currently being recognized under the name of another much more common species. Such cryptic species are common in crayfish biology and other fields of invertebrate biology. For example, what is currently known as a single species of crayfish may range from Canada to Alabama, but when you look closely at their morphology (and genetics), there may be multiple, or even dozens of, species hidden within this "species complex" that await formal description.

Species complexes form large-scale taxonomic puzzles and have proved to be a substantial undertaking by daring astacologists willing to try to solve them. Several species complexes are well known to crayfish biologists and will likely be split into several species in the coming years and decades.

The Common Crayfish (*Cambarus bartonii*) is one of many crayfishes with a large geographic range and varies widely depending on where it is caught. Focused taxonomic work on this species will likely reveal that there are several undescribed species, to be described in the future. Photo by Guenter Schuster.

A Banded Mudbug (*Lacunicambarus freudensteini*). Photo by Guenter Schuster.

WITH GLON'S DISCOVERY, the Banded Mudbug now has an official place in the long list of Mississippi's and Alabama's biodiversity. Prior to this work, this crayfish might have gone undetected or been mislabeled as a Painted Devil Crayfish.

Now, Glon's findings have elevated a taxonomic conundrum into a matter of conservation. As the species description makes clear, the Banded Mudbug is a scarce creature, known to occur in only nine locations. This is not a common mudbug; it is a rare gem, restricted in both distribution and abundance.

Crayfish are well supported in both of the states where the Banded Mudbug is found, and it didn't take long for this species to be listed as critically imperiled by state agencies in both Mississippi and Alabama. This designation acknowledges the species' value and may prevent potential harm that could come its way. Being critically imperiled means that the states want to protect the species from construction or agriculture that would alter the few roadside ditch habitats where it is found. Such state-level protection is also a step toward gaining federal attention and possibly being considered for listing under the Endangered Species Act.

The Banded Mudbug is truly a one-of-a-kind creature, not just because of its vibrant turquoise and orange banding but also because of the dangers it faces. Glon and his colleague Susie Adams, a Mississippi-based crayfish biologist, aren't worried about invasive species or habitat destruction in the quiet rural area where the mudbug lives. Their main concern is the potential rise in sea level.

According to some climate models, a sea-level rise of 2.7 meters (8.85 ft) is expected in the future, which could inundate at least two of the nine known locations of the Banded Mudbug. Not only could the rising seawater destroy these populations, but the seawater could also penetrate underground and impact this species and other nearby burrowing crayfish. Although crayfish can tolerate salt water for short periods, it is unknown whether the Banded Mudbug or nearby crayfishes have this ability. On top of this, a single hurricane could also have a devastating effect on these crayfish, potentially knocking out many of the known populations.

THE DISCOVERY OF THE BANDED MUDBUG is a testament to the importance of crayfish taxonomy and the role it plays in conservation. Had it not been for Glon's expertise and his ability to recognize the unique characteristics of this species, the Banded Mudbug would still be living in relative obscurity, just another Painted Devil Crayfish among many.

But now, with a name and a place in the scientific record, the Banded Mudbug has a fighting chance. It has the opportunity to be protected and to have its unique habitat and environment safeguarded against the threat of rising sea levels or any other potential disasters. And who knows what other species are out there, waiting to be discovered, waiting to have their names given, and waiting to be protected. The Banded Mudbug is just one of many examples of the incredible diversity of life on our planet and the importance of science in preserving it for future generations.

Unfortunately, the science of taxonomy is in serious decline. Funding agencies want to pump money into big, sexy projects that give concrete benefits to the environment and not a quirky biologist who wants to distinguish one mudbug from another. Funding for taxonomy work is at an all-time low, and now the entire world of North American crayfish taxonomy is contained within the brains of a half dozen or so individuals, many of whom are in the end stages of their careers. Zac Loughman commonly preaches that "crayfish biologists and taxonomists are themselves, an endangered species." The state of taxonomy is similarly bleak in other fields.

Without these specialists, countless species will go undiscovered, and their unique quirks and characteristics may never be appreciated by the wider world. It's a grievous thought that the survival of an entire species could rest on the shoulders of just a few individuals, tasked with the unenviable task of navigating the complex world of taxonomy and species description. Next time you come across a seemingly unremarkable animal, consider the years of expertise that may have gone into identifying it and the tireless work of those dedicated to preserving not only the species itself but the knowledge that allows us to appreciate their existence. As the story of the Banded Mudbug has shown, you never know what fascinating secrets lie beneath the surface, waiting to be uncovered by a dedicated taxonomist.

I am not petitioning for everyone reading this book to become a crayfish taxonomist. This area of biology can be challenging, but it is important nonetheless, particularly when it comes to groups like crayfishes. There are still many undescribed species out there, waiting to be discovered and named. A lack of funding and job opportunities has led many skilled crayfish taxonomists and biologists to seek employment elsewhere, in fields that receive more consistent state and federal support, such as fish or mammals, which generate income through hunting. Nevertheless, it is essential to recognize the crucial role that crayfish play in supporting the ecosystems that many of these other animals rely on.

# ENDANGERED SPECIES

As of June 2023, there are ten North American crayfish species protected by the Endangered Species Act. Six species are listed as endangered, and four are listed as threatened. In the coming years, this number is expected to grow at an unprecedented rate; many additional species are being petitioned to receive protection under the ESA.

When a crayfish species become listed under the ESA, it's more likely to be protected, because the act guarantees funds dedicated to its conservation, which ultimately leads to support for future crayfish biologists. In the time it took me to write this book alone (2021–24), three crayfish species were added to the list of federally protected species, taking the number from seven to ten.

The two most recent crayfishes protected by the ESA are the Big Creek Crayfish (Faxonius peruncus) and the St. Francis River Crayfish (Faxonius quadruncus), which are both listed as threatened. Mining and sedimentation have reduced their range down to just a few streams in Missouri, where they are being invaded by Woodland Crayfish (Faxonius hylas).

A Big Creek Crayfish (*Faxonius peruncus*). Photo by Chris Lukhaup.

A St. Francis River Crayfish (*Faxonius quadruncus*). Photo by Chris Lukhaup.

# Our Native Crustaceans

If someone sees me fidgeting around in a creek or rummaging in a roadside burrow, their curiosity is usually piqued. No matter where I am or whether they know the creatures I'm after as crayfish, crawfish, crawdads, or even mudbugs, they're always interested when I tell them that I spend my life thinking about these critters. It often elicits a mixture of nostalgia and surprise. People are amazed to learn that hundreds of individuals across the United States, and even more globally, spend their lives flipping rocks and splashing in the mud, just as they did when they were children. Although these animals are not often seen in plain sight and are rarely discussed outside of the context of being boiled and consumed in large quantities, they are widely known. However, few people are aware of the impact that crayfish have on our lives (beyond being served up with butter).

If there's one thing I've come to appreciate about crayfish, it's their resilience and adaptability. They have spread throughout the United States and beyond, diversifying into over 700 species, and can be collected in areas that are inhabitable by others. In streams, lakes, rivers, and ponds, crayfish with large, cumbersome claws are often abundant. At water's edge, burrowing species may be present in puzzlingly bright colors that remain a mystery to the biologists that study them. Other crayfish have secluded themselves in caves so far away from light, and for such a long time, that they have entirely lost their eyes. Some species only occur in a single river basin, whereas others span a geographic range across dozens of states.

Wherever crayfish are, their ecological role is apparent. Anything smaller than them is a potential dinner item, whether it's a plant, insect, mammal, or fish. Bigger predators relish opportunities to feast on crayfish, whether it be a raccoon flipping rocks in a creek bed or an owl diving onto an exposed blue burrower. Some animals have such a close relationship with crayfish that if the crayfish disappear, so will they. And the burrows that crayfish construct are used by many, including a hefty share of species that are themselves of conservation concern. Crayfish provide a strong support system for both aquatic and terrestrial habitats by acting as ecosystem engineers and keystone species.

But many crayfish are on the decline. Nearly 50% of all species require conservation attention. Many of the remaining species are considered data deficient, and sadly, astacologists don't even know enough about them to make a statement as to how they are doing. The top threats in the United States come from the human-mediated spread of invasive species, as well as the continuing impacts of humans on the chemical and physical makeup of our freshwater ecosystems. Some crayfish species are perishing before they are even described, and they may be wiped out before they even receive a name. Although crayfish protection is probably at an all-time high, these efforts can be interpreted as last-ditch efforts rather than best-case scenarios.

Crayfish deserve the chance to do what they have done for millions of years, because the freshwater environments we cherish most are intimately connected to, in one way or another, crayfish, crawfish, and crawdads. They deserve the opportunity to burrow under a rock or in the mud. They deserve the opportunity to eat—and get eaten by others. But we have to make sure that they have this opportunity by providing them with healthy and sustainable environments. Because when the crayfish disappear, other species quickly follow.

Despite the lingering fears, I remain hopeful for the future of crayfish, knowing that with the right efforts, they will continue their journey in the waters they call home. My role in this journey is to continue doing what I have always done, by flipping every rock that might hide a crayfish beneath it and by bringing up crayfish in every conversation no matter how unrelated it may be (grocery store cashiers and Uber drivers, beware). What started out as a childhood obsession is now my career. I spend my days exploring the diversity of crayfishes, a privilege that allows me to engage with these creatures on a daily basis. As long as there are rocks to flip and burrows to dig, I know there is always more to learn.

# Acknowledgments

THE HARDEST PART of writing this book was deciding what to include and what to exclude. I was forced to rely on my own experiences and the experiences of my closest colleagues. This means that many stories, species, and deserving biologists had to be left out. Be aware that this book could be written five times again, with each version telling different stories.

On that note, I would like to thank all of the crayfish biologists who opened up their lives to me. Learning about crayfish from the "crayfish people" was my favorite part about writing this book. In no particular order, I thank Chris Taylor, Guenter Schuster, Jim Stoeckel, Keith Crandall, Jim Fetzner, Chris Lukhaup, Chris Bonvillian, Eric Larson, Jens Herberholz, Susie Adams, Bob DiStefano, Paul Moore, Mael Glon, Lindsey Reisinger, Zen Faulkes, Alex Palaoro, Roger Thoma, Michael Lannoo, Frank Krasne, Thomas Breithaupt, and Bronwyn Williams. Special thanks to Maria Ellis, Jeff Cook, and Koen Breedveld for taking me in and giving me a glimpse into their lives protecting the Shasta Crayfish. I also need to express my gratitude to Zac Loughman, who has served as a boss, a mentor, a colleague, and a friend throughout my development as a crayfish biologist. Zac has provided overwhelming support since we started working together.

One of the biggest motivators in writing this book was to showcase crayfish diversity. This could not be possible without the photographers who graciously provided photos. These individuals have spent many hours in the streams and in the mud photographing crayfish. This book would not be the same without their contribution. Guenter Schuster and Chris Lukhaup have provided a large chunk of these photos, and I appreciate their generosity.

Several others played key roles throughout this process. I thank Ben Goldfarb, the author of a similar narrative-driven book on beavers. I reached out to him at random for help in this process, and he was kind enough to respond. I also thank Josh Nolte. Without Josh's help, I would not be the writer that I am today. Out of sheer interest and friendship, Josh edited most of this book. He fixed countless grammatical inconsistencies while ensuring that my writing was whimsical and worth reading. I also thank my editor, Lucas Church, for providing me with comments throughout the book and other members of UNC Press who have guided me over the past few years.

Lastly, I thank my friends and family for understanding my early morning writing rituals, which often forced me to bed before sunset. Without their support, I would not have had the initial animal experiences that led me down this path.

# Bibliography

## CHAPTER 1

Crandall, Keith A., and Sammy DeGrave. "An Updated Classification of the Freshwater Crayfishes (Decapoda: Astacidea) of the World, with a Complete Species List." *Journal of Crustacean Biology* 37, no. 5 (2017): 615–53.

Hart, C. W., Jr. *A Dictionary of Non-Scientific Names of Freshwater Crayfishes (Astacoidea and Parastacoidea), Including Other Words and Phrases Incorporating Crayfish Names.* Smithsonian Contributions to Anthropology 38. Smithsonian Institution Press, 1994. https://doi.org/10.5479/si.00810223.38.1.

"Horton H. Hobbs, Jr. (29 March 1914–22 March 1994)." *Journal of Crustacean Biology* 15, no. 4 (1995): 797–99.

Huxley, Thomas Henry. *The Crayfish: An Introduction to the Study of Zoology.* International Scientific Series 28. C. Kegan Paul, 1880.

Reynolds, Julian, Catherine Souty-Grosset, and Alastair Richardson. "Ecological Roles of Crayfish in Freshwater and Terrestrial Habitats." *Freshwater Crayfish* 19, no. 2 (2013): 197–218.

Richman, Nadia I., Monika Böhm, Susan B. Adams, et al. "Multiple Drivers of Decline in the Global Status of Freshwater Crayfish (Decapoda: Astacidea)." *Philosophical Transactions of the Royal Society B: Biological Sciences* 370, no. 1662 (2015): 1–11.

Schuster, Guenter A. "Review of Crayfish Color Patterns in the Family Cambaridae (Astacoidea), with Discussion of Their Possible Importance." *Zootaxa* 4577, no. 1 (2020): 63–98.

Schuster, Guenter A., Christopher A. Taylor, and Stuart W. McGregor. *Crayfishes of Alabama.* University of Alabama Press, 2022.

Taylor, Christopher A., Robert J. DiStefano, Eric R. Larson, and James A. Stoeckel. "Towards a Cohesive Strategy for the Conservation of the United States' Diverse and Highly Endemic Crayfish Fauna." *Hydrobiologia* 846, no. 1 (2019): 39–58.

Taylor, Christopher A., and Guenter A. Schuster. *The Crayfishes of Kentucky.* Illinois Natural History Survey, 2004.

Taylor, Christopher A., Guenter A. Schuster, John E. Cooper, et al. "A Reassessment of the Conservation Status of Crayfishes of the United States and Canada after 10+ Years of Increased Awareness." *Fisheries* 32, no. 8 (2007): 372–89.

## CHAPTER 2

Aquiloni, Laura, and Francesca Gherardi. "Extended Mother-Offspring Relationships in Crayfish: The Return Behaviour of Juvenile *Procambarus clarkii*." *Ethology* 114, no. 10 (2008): 946–54.

Berrill, Michael, and Brian Chenoweth. "The Burrowing Ability of Nonburrowing Crayfish." *American Midland Naturalist* 108, no. 1 (1982): 199–201.

Cuthill, Innes C., William L. Allen, Kevin Arbuckle, et al. "The Biology of Color." *Science* 357, no. 6350 (2017). https://doi.org/10.1126/science.aan0221.

Dalosto, Marcelo M., A. V. Palaoro, and S. Santos. "Mother-Offspring Relationship in the Neotropical Burrowing Crayfish *Parastacus pilimanus* (Von Martens, 1869) (Decapoda, Parastacidae)." *Crustaceana* 85, no. 11 (2012): 1305–15.

Graham, Zackary A. "Prevalence and Potential Evolutionary Significance of Color Variants in Freshwater Crayfishes (Decapoda: Astacidea)." *Journal of Crustacean Biology* 43, no. 3 (2023): 1–5.

Graham, Zackary A., and Dylan J. Padilla Perez. "Correlated Evolution of Conspicuous Coloration and Burrowing in Crayfish." *Proceedings of the Royal Society B: Biological Sciences* 291, no. 2026 (2024). https://doi.org/10.1101/2023.07.03.547601.

Hobbs, H. H. *The Crayfishes of Florida*. Biological Science Series 3, no. 2. University of Florida, 1942.

Hobbs, H. H. *The Crayfishes of Georgia*. Smithsonian Contributions to Zoology 318. Smithsonian Institution Press, 1981. https://doi.org/10.5479/si.00810282.318.

Kouba, Antonin, Jan Tikal, Petr Cisar, et al. "The Significance of Droughts for Hyporheic Dwellers: Evidence from Freshwater Crayfish." *Scientific Reports*, no. 6 (2016): 1–7.

Norrocky, M. James. "Observations on the Ecology, Reproduction and Growth of the Burrowing Crayfish *Fallicambarus* (*Creaserinus*) *fodiens* (Decapoda: Cambaridae) in North-Central Ohio." *American Midland Naturalist* 125, no. 1 (1991): 75.

Ortmann, Arnold E. *The Crawfishes of the State of Pennsylvania*. Memoirs of the Carnegie Museum 2, no. 10. Board of Trustees of the Carnegie Institute, 1906. https://doi.org/10.5962/bhl.title.10407.

Richardson, Alastair M. M. "Behavioral Ecology of Semiterrestrial Crayfish." In *Evolutionary Ecology of Social and Sexual Systems: Crustaceans as Model Organisms*, edited by J. Emmett Duffy and Martin Thiel, 219–338. Oxford University Press, 2007.

Richardson, Alastair M. M. "The Effect of the Burrows of a Crayfish on the Respiration of the Surrounding Soil." *Soil Biology and Biochemistry* 15, no. 3 (1983): 239–42.

Schuster, Guenter A. "Review of Crayfish Color Patterns in the Family Cambaridae (Astacoidea), with Discussion of Their Possible Importance." *Zootaxa* 4577, no. 1 (2020): 63–98.

Welch, Shane M., and Arnold G. Eversole. "The Occurrence of Primary Burrowing Crayfish in Terrestrial Habitat." *Biological Conservation* 130, no. 3 (2006): 458–64.

## CHAPTER 3

Bavetz, Mark. "Geographic Variation, Status, and Distribution of Kirtland's Snake (*Clonophis kirtlandii* Kennicott) in Illinois." *Transactions of the Illinois State Academy of Science* 87, no. 3 (1994): 151–63.

Berrill, Michael, and Brian Chenoweth. "The Burrowing Ability of Nonburrowing Crayfish." *American Midland Naturalist* 108, no. 1 (1982): 199–201.

Creed, Robert P., James Skelton, Kaitlin J. Farrell, and Bryan L. Brown. "Strong Effects of a Mutualism on Freshwater Community Structure." *Ecology* 102, no. 2 (2021). https://doi.org/10.1002/ecy.3225.

DiStefano, Robert J. "Trophic Interactions between Missouri Ozarks Stream Crayfish Communities and Sport Fish Predators: Increased Abundance and Size Structure of Predators Cause Little Change in Crayfish Community Density." Final Report, Project F-1-R-54, Study S-41, Job 4. Missouri Department of Conservation, Federal Aid in Sport Fish Restoration, 2005.

Engbrecht, Nathan, and Michael Lannoo. "A Review of the Status and Distribution of Crawfish Frogs (*Lithobates areolatus*) in Indiana." *Proceedings of the Indiana Academy of Science* 119, no. 1 (2010): 64–73.

Glon, Mael G., and Roger F. Thoma. "An Observation of the Use of Devil Crayfish (*Cambarus* cf. *diogenes*) Burrows as Brooding Habitat by Eastern Cicada Killer Wasps (*Sphecius speciosus*)." *Freshwater Crayfish* 23, no. 1 (2017): 55–57.

Goldfarb, Ben. *Eager: The Surprising, Secret Live of Beavers and Why They Matter.* Chelsea Green, 2018.

Graham, Zackary A., and Zachary J. Loughman. "Natural History and Ecology of the Slender Crayfish (*Faxonius compressus*): An Ecosystem Engineer in the Western Highland Rim, USA." *Journal of Natural History* 21–24, no. 57 (2023): 1235–56.

Hoffman, Andrew S., Jennifer L. Heemeyer, Perry J. Williams, et al. "Strong Site Fidelity and a Variety of Imaging Techniques Reveal Around-the-Clock and Extended Activity Patterns in Crawfish Frogs (*Lithobates areolatus*)." *BioScience* 60, no. 10 (2010): 829–34.

Holycross, Andrew T., and Stephen P. Mackessy. "Variation in the Diet of *Sistrurus catenatus* (Massasauga), with Emphasis on *Sistrurus catenatus edwardsii* (Desert Massasauga)." *Journal of Herpetology* 36, no. 3 (2002): 454–64.

Jackrel, Sara L., and Howard K. Reinert. "Behavioral Responses of a Dietary Specialist, the Queen Snake (*Regina septemvittata*), to Potential Chemoattractants Released by Its Prey." *Journal of Herpetology* 45, no. 3 (2011): 272–76.

Krupa, James J., Kevin R. Hopper, and Monica A. Nguyen. "Dependence of the Dwarf Sundew (*Drosera brevifolia*) on Burrowing Crayfish Disturbance." *Plant Ecology* 222, no. 4 (2021): 459–67.

Pintor, Lauren M., and Daniel A. Soluk. "Evaluating the Non-Consumptive, Positive Effects of a Predator in the Persistence of an Endangered Species." *Biological Conservation* 130, no. 4 (2006): 584–91.

Powell, Sylvia D., Thomas A. Gorman, and Carola A. Haas. "Nightly Movements and Use of Burrows by Reticulated Flatwoods Salamanders (*Ambystoma bishopi*) in Breeding Wetlands." *Florida Scientist* 78, no. 3 (2015): 149–55.

Reynolds, Julian, Catherine Souty-Grosset, and Alastair Richardson. "Ecological Roles of Crayfish in Freshwater and Terrestrial Habitats." *Freshwater Crayfish* 19, no. 2 (2013): 197–218.

Roth, Brian M., Catherine L. Hein, and M. Jake Vander Zanden. "Using Bioenergetics and Stable Isotopes to Assess the Trophic Role of Rusty Crayfish (*Orconectes rusticus*) in Lake Littoral Zones." *Canadian Journal of Fisheries and Aquatic Sciences* 63, no. 2 (2006): 335–44.

Skelton, James, Kaitlin J. Farrell, Robert P. Creed, et al. "Servants, Scoundrels, and Hitchhikers: Current Understanding of the Complex Interactions between Crayfish and Their Ectosymbiotic Worms (Branchiobdellida)." *Freshwater Science* 32, no. 4 (2013): 1345–57.

Statzner, Bernhard, O. Peltret, and S. Tomanova. "Crayfish as Geomorphic Agents and Ecosystem Engineers: Effect of a Biomass Gradient on Baseflow and Flood-Induced Transport of Gravel and Sand in Experimental Streams." *Freshwater Biology* 48, no. 1 (2003): 147–63.

Stites, Andrew J., Christopher A. Taylor, and Ethan J. Kessler. "Trophic Ecology of the North American Crayfish Genus *Barbicambarus* Hobbs, 1969 (Decapoda: Astacoidea: Cambaridae): Evidence for a Unique Relationship between Body Size and Trophic Position." *Journal of Crustacean Biology* 37, no. 3 (2017): 263–71.

Tran, Mark V., and Amy Manning. "Seasonal Diet Shifts in the Rusty Crayfish, *Faxonius rusticus* (Girard)." *Ohio Biological Survey Notes*, no. 9 (2019): 41–45.

Tumlison, Renn, and Kory G. Roberts. "Prey-Handling Behavior in the Gulf Crayfish Snake (*Liodytes rigida*)." *Herpetological Conservation and Biology* 13, no. 3 (2018): 617–21.

Usio, Nisikawa, and Colin R. Townsend. "Roles of Crayfish: Consequences of Predation and Bioturbation for Stream Invertebrates." *Ecology* 85, no. 3 (2004): 807–22.

Williams, Bronwyn W., and Patricia G. Weaver. "Historical Review of the Taxonomy and Classification of Entocytheridae (Crustacea: Ostracoda: Podocopida)." *Zootaxa* 4448, no. 1 (2018). https://doi.org/10.11646/zootaxa.4448.1.1.

Williams, Perry J., Joseph R. Robb, and Daryl R. Karns. "Occupancy Dynamics of Breeding Crawfish Frogs in Southeastern Indiana." *Wildlife Society Bulletin* 36, no. 2 (2012): 350–57.

Wright, Justin P., Clive G. Jones, and Alexander S. Flecker. "An Ecosystem Engineer, the Beaver, Increases Species Richness at the Landscape Scale." *Oecologia* 132, no. 1 (2002): 96–101.

## CHAPTER 4

Bellman, Kirstie L., and Franklin B. Krasne. "Adaptive Complexity of Interactions between Feeding and Escape in Crayfish." *Science* 221, no. 4612 (1983): 779–81.

Berrill, Michael, and Michael Arsenault. "The Breeding Behaviour of a Northern Temperate Orconectid Crayfish, *Orconectes rusticus*." *Animal Behaviour* 32, no. 2 (1984): 333–39.

Berrill, Michael, and Michael Arsenault. "Spring Breeding of a Northern Temperate Crayfish, *Orconectes rusticus*." *Canadian Journal of Zoology* 60, no. 11 (1982): 2641–45.

Brody, Debra J., and Qiuping Gu. "Antidepressant Use among Adults: United States, 2015–2018." *NCHS Data Brief*, no. 377 (September 2020): 1–8.

Crandall, Keith A. "Crayfish as Model Organisms." *Freshwater Crayfish* 13, no. 1 (2002): 3–10.

Edwards, Donald H., William J. Heitler, and Franklin B. Krasne. "Fifty Years of a Command Neuron: The Neurobiology of Escape Behavior in the Crayfish." *Trends in Neurosciences* 22, no. 4 (1999): 153–61.

Edwards, Donald H., Fadi A. Issa, and Jens Herberholz. "The Neural Basis of Dominance Hierarchy Formation in Crayfish." *Microscopy Research and Technique* 60, no. 3 (2003): 369–76.

Herberholz, Jens, Brian L. Antonsen, and Donald H. Edwards. "A Lateral Excitatory Network in the Escape Circuit of Crayfish." *Journal of Neuroscience* 22, no. 20 (2002): 9078–85.

Herberholz, Jens, Fadi A. Issa, and Donald H. Edwards. "Patterns of Neural Circuit Activation and Behavior during Dominance Hierarchy Formation in Freely Behaving Crayfish." *Journal of Neuroscience* 21, no. 8 (2001): 2759–67.

Herberholz, Jens, and Gregory D. Marquart. "Decision Making and Behavioral Choice during Predator Avoidance." *Frontiers in Neuroscience*, no. 6 (2012): 1–15.

Herberholz, Jens, Marjorie M. Sen, and Donald H. Edwards. "Escape Behavior and Escape Circuit Activation in Juvenile Crayfish during Prey-Predator Interactions." *Journal of Experimental Biology* 207, no. 11 (2004): 1855–63.

Huxley, Thomas Henry. *The Crayfish: An Introduction to the Study of Zoology*. International Scientific Series 28. C. Kegan Paul, 1880.

Reisinger, Alexander J., Lindsey S. Reisinger, Erinn K. Richmond, and Emma J. Rosi. "Exposure to a Common Antidepressant Alters Crayfish Behavior and Has Potential Subsequent Ecosystem Impacts." *Ecosphere* 12, no. 6 (2021). https://doi.org/10.1002/ecs2.3527.

Swierzbinski, Matthew E., Andrew R. Lazarchik, and Jens Herberholz. "Prior Social Experience Affects the Behavioral and Neural Responses to Acute Alcohol in Juvenile Crayfish." *Journal of Experimental Biology* 220, no. 8 (2017): 1516–23.

Wiersma, C. A. G. "Giant Nerve Fiber System of the Crayfish; a Contribution to Comparative Physiology of Synapse." *Journal of Neurophysiology* 10, no. 1 (1947): 23–38.

Wine, J. J., and F. B. Krasne. "The Organization of Escape Behaviour in the Crayfish." *Journal of Experimental Biology* 56, no. 1 (1972): 1–18.

## CHAPTER 5

Basil, Jennifer, and David Sandeman. "Crayfish (*Cherax destructor*) Use Tactile Cues to Detect and Learn Topographical Changes in Their Environment." *Ethology* 106, no. 3 (2000): 247–59.

Belanger, Rachelle, and Paul A. Moore. "A Comparative Analysis of Setae on the Pereiopods of Reproductive Male and Female *Orconectes rusticus* (Decapoda: Astacidae)." *Journal of Crustacean Biology* 33, no. 3 (2013): 309–16.

Belanger, Rachelle, Xiang Ren, Katherine McDowell, Steven Chang, Paul Moore, and Barbara Zielinski. "Sensory Setae on the Major Chelae of Male Crayfish, *Orconectes* rusticus (Decapoda: Astacidae)—Impact of Reproductive State on Function and Distribution." *Journal of Crustacean Biology* 28, no. 1 (2008): 27–36.

Bergman, Daniel A., Arthur L. Martin, and Paul A. Moore. "Control of Information Flow through the Influence of Mechanical and Chemical Signals during Agonistic Encounters by the Crayfish, *Orconectes rusticus*." *Animal Behaviour* 70, no. 3 (2005): 485–96.

Bergman, Daniel A., Christopher N. Redman, Kandice C. Fero, Jodie L. Simon, and Paul A. Moore. "The Impacts of Flow on Chemical Communication Strategies and Fight Dynamics of Crayfish." *Marine and Freshwater Behaviour and Physiology* 39, no. 4 (2006): 245–58.

Bovbjerg, Richard V. "Dominance Order in the Crayfish *Orconectes virilis* (Hagen)." *Physiological Zoology* 26, no. 2 (1953): 173–78.

Bovbjerg, Ricard V. "Some Factors Affecting Aggressive Behavior in Crayfish." *Physiological Zoology* 29, no. 2 (1956): 127–36.

Breithaupt, Thomas. "Fan Organs of Crayfish Enhance Chemical Information Flow." *Biological Bulletin* 200, no. 2 (2001): 150–54.

Breithaupt, Thomas. "Sound Perception in Aquatic Crustaceans." In *The Crustacean Nervous System*, edited by Konrad Wiese, 548–58. Springer, 2002.

Breithaupt, Thomas, and Petra Eger. "Urine Makes the Difference: Chemical Communication in Fighting Crayfish Made Visible." *Journal of Experimental Biology* 205, no. 9 (2002): 1221–31.

Breithaupt, Thomas, Daniel P. Lindstrom, and Jelle Atema. "Urine Release in Freely Moving Catheterised Lobsters (*Homarus americanus*) with Reference to Feeding and Social Activities." *Journal of Experimental Biology* 202, no. 7 (1999): 837–44.

Caves, Eleanor M., Stephen Nowicki, and Sönke Johnsen. "Von Uexküll Revisited: Addressing Human Biases in the Study of Animal Perception." *Integrative and Comparative Biology* 59, no. 6 (2019): 1451–62.

Chiandetti, C., and A. Caputi. "Visual Shape Recognition in Crayfish as Revealed by Habituation." *Animal Behavior and Cognition* 4, no. 3 (2017): 242–51.

Dearborn, George V. N. "Notes on the Individual Psychophysiology of the Crayfish." *American Journal of Physiology* 3, no. 9 (1900): 404–33.

DeForest, Mellon. "Smelling, Feeling, Tasting and Touching: Behavioral and Neural Integration of Antennular Chemosensory and Mechanosensory Inputs in the Crayfish." *Journal of Experimental Biology* 215, no. 13 (2012): 2163–72.

Favaro, Livio, Tina Tirelli, Marco Gamba, and Daniela Pessani. "Sound Production in the Red Swamp Crayfish *Procambarus clarkii* (Decapoda: Cambaridae)." *Zoologischer Anzeiger* 250, no. 2 (2011): 143–50.

Gherardi, Francesca, and Laura Aquiloni. "Sexual Selection in Crayfish: A Review." In *New Frontiers in Crustacean Biology: Proceedings of the TCS Summer Meeting, Tokyo, 20–24 September, 2009*, edited by Akira Asakura, 213–23. Crustuceana Monographs 15. Brill, 2011. https://brill.com/display/book/edcoll/9789047427711/Bej.9789004174252.i-354_019.xml.

Hobbs, H. H., and Stanley A. Rewolinski. "Notes on the Burrowing Crayfish *Procambarus* (*Girardiella*) *gracilis* (Bundy) (Decapoda, Cambaridae) from Southeastern Wisconsin, U.S.A." *Crustaceana* 48, no. 1 (1985): 26–33.

Mantel, Linda H., and Linda L. Farmer. "Osmotic and Ionic Regulation." In *Internal Anatomy and Physiological Regulation*, edited by Linda H. Mantel, 53–161. Vol. 5 of *The Biology of Crustacea*, edited by Dorothy E. Bliss. Academic Press, 1983.

McLay, Colin L., and Anekke M. van den Brink. "Crayfish Growth and Reproduction." In *Biology and Ecology of Crayfish*, edited by Matt Longshaw and Paul Stebbing, 62–116. CRC, 2016.

Moore, Paul A. "Agonistic Behavior in Freshwater Crayfish: The Influence of Intrinsic and Extrinsic Factors on Aggressive Behavior and Dominance." In *Evolutionary Ecology of Social and Sexual Systems: Crustaceans as Model Organisms*, edited by J. Emmett Duffy and Martin Thiel, 90–114. Oxford University Press, 2007.

Moore, Paul A., and Daniel A. Bergman. "The Smell of Success and Failure: The Role of Intrinsic and Extrinsic Chemical Signals on the Social Behavior of Crayfish." *Integrative and Comparative Biology* 45, no. 4 (2005): 650–57.

Sandeman, David C. "Physical Properties, Sensory Receptors and Tactile Reflexes of the Antenna of the Australian Freshwater Crayfish *Cherax destructor*." *Journal of Experimental Biology* 141, no. 1 (1989): 197–217.

Sandeman, David C., and Lon A. Wilkens. "Sound Production by Abdominal Stridulation in the Australian Murray River Crayfish, *Euastacus armatus*." *Journal of Experimental Biology* 99, no. 1 (1982): 469–72.

Stein, Roy A. "Sexual Dimorphism in Crayfish Chelae: Functional Significance Linked to Reproductive Activities." *Canadian Journal of Zoology* 54, no. 2 (1976): 220–27.

Tautz, J., and David C. Sandeman. "The Detection of Waterborne Vibration by Sensory Hairs on the Chelae of the Crayfish." *Journal of Experimental Biology* 88, no. 1 (1980): 351–56.

Vogt, Günter, Laura Tolley, and Gerhard Scholtz. "Life Stages and Reproductive Components of the Marmorkrebs (Marbled Crayfish), the First Parthenogenetic Decapod Crustacean." *Journal of Morphology* 261, no. 3 (2004): 286–311.

Wald, G. "Single and Multiple Visual Systems in Arthropods." *Journal of General Physiology* 51, no. 2 (1968): 125–56.

Wald, G. "Visual Pigments of Crayfish." *Nature* 215, no. 5106 (1967): 615–16.

Weagle, K. V., and G. W. Ozburn. "Sexual Dimorphism in the Chela of *Orconectes virilis* (Hagen)." *Canadian Journal of Zoology* 48, no. 5 (1970): 1041–42.

## CHAPTER 6

Anastacio, Pedro M., Ferreira, Miriam P., Banha, Filipe, et al. "Waterbird-Mediated Passive Dispersal Is a Viable Process for Crayfish (*Procambarus clarkii*)." *Aquatic Ecology*, no. 45 (2014): 1–10.

Aquiloni, Laura, Aldo Becciolini, Roberto Berti, Sauro Porciani, Carmen Trunfio, and Francesca Gherardi. "Managing Invasive Crayfish: Use of X-Ray Sterilisation of Males." *Freshwater Biology* 54, no. 7 (2009): 1510–19.

Buktenica, M. W., S. F. Girdner, A. M. Ray, D. K. Hering, and J. Umek. "The Impact of Introduced Crayfish on a Unique Population of Salamander in Crater Lake, Oregon." *Park Science* 32, no. 1 (2015): 5–12.

Chucholl, Christoph. "The Bad and the Super-Bad: Prioritising the Threat of Six Invasive Alien to Three Imperilled Native Crayfishes." *Biological Invasions* 18, no. 7 (2016): 1967–88.

Chucholl, Christoph. "Invaders for Sale: Trade and Determinants of Introduction of Ornamental Freshwater Crayfish." *Biological Invasions* 15, no. 1 (2013): 125–41.

Faulkes, Zen. "The Global Trade in Crayfish as Pets." *Crustacean Research*, no. 44 (2015): 75–92.

Faulkes, Zen. "Marmorkrebs (*Procambarus fallax* f. *virginalis*) Are the Most Popular Crayfish in the North American Pet Trade." *Knowledge and Management of Aquatic Ecosystems* 416, no. 20 (2015): 1–15.

Faulkes, Zen. "Prohibiting Pet Crayfish Does Not Consistently Reduce Their Availability Online." *Nauplius*, no. 26 (November 2018): 1–11.

Gherardi, Francesca, Laura Aquiloni, Javier Diéguez-Uribeondo, and Elena Tricarico. "Managing Invasive Crayfish: Is There a Hope?" *Aquatic Sciences* 73, no. 2 (2011): 185–200.

Gutekunst, Julian, Olena Maiakovska, Katharina Hanna, et al. "Phylogeographic Reconstruction of the Marbled Crayfish Origin." *Communications Biology* 4, no. 1 (2021): 1–6.

Huner, Jay V. *Freshwater Crayfish Aquaculture in North America, Europe, and Australia: Families Astacidae.* CRC, 1994.

Irwin, Sam. *Louisiana Crawfish: A Succulent History of the Cajun Crustacean.* Arcadia, 2008.

Jimenez, Stephanie A., and Zen Faulkes. "Can the Parthenogenetic Marbled Crayfish Marmorkrebs Compete with Other Crayfish Species in Fights?" *Journal of Ethology* 29, no. 1 (2011): 115–20.

Keller, Reuben P., Kristin Frang, and David M. Lodge. "Preventing the Spread of Invasive Species: Economic Benefits of Intervention Guided by Ecological Predictions." *Conservation Biology* 22, no. 1 (2008): 80–88.

Larson, Eric R., Timothy A. Kreps, Brett Peters, Jody A. Peters, and David M. Lodge. "Habitat Explains Patterns of Population Decline for an Invasive Crayfish." *Ecology* 100, no. 5 (2019): 1–7.

Larson, Eric R., and Julian D. Olden. "Do Schools and Golf Courses Represent Emerging Pathways for Crayfish Invasions?" *Aquatic Invasions* 3, no. 4 (2008): 465–68.

Lodge, David M., Roya Stein, Kenneth M. Brown, et al. "Predicting Impact of Freshwater Exotic Species on Native Biodiversity: Challenges in Spatial Scaling." *Austral Ecology* 23, no. 1 (1998) 53–67.

Maiakovska, Olena, Ranja Andriantsoa, Sina Tönges, et al. "Genome Analysis of the Monoclonal Marbled Crayfish Reveals Genetic Separation over a Short Evolutionary Timescale." *Communications Biology* 4, no. 1 (2021): 1–7.

Penn, George H. "A Study of the Life History of the Louisiana Red-Crawfish, *Cambarus clarkii* Girard." *Ecology* 24, no. 1 (1943): 1–18.

Pitre, Len. *The Crawfish Book.* University of Mississippi Press, 2010.

Thomas, John Rhidian, Stephanie Masefield, Rhiannon Hunt, et al. "Terrestrial Emigration Behaviour of Two Invasive Crayfish Species." *Behavioural Processes*, no. 167 (October 2019): 10–13.

Vogt, G. "The Marbled Crayfish: A New Model Organism for Research on Development, Epigenetics and Evolutionary Biology." *Journal of Zoology* 276, no. 1 (2008): 1–13.

Wilson, Karen A., John J. Magnuson, David M. Lodge, et al. "A Long-Term Rusty Crayfish (*Orconectes rusticus*) Invasion: Dispersal Patterns and Community Change in a North Temperate Lake." *Canadian Journal of Fisheries and Aquatic Sciences* 61, no. 11 (2004): 2255–66.


## CHAPTER 7

Cowart, Dominique A., Koen G. H. Breedveld, Maria J. Ellis, Joshua M. Hull, and Eric R. Larson. "Environmental DNA (EDNA) Applications for the Conservation of Imperiled Crayfish (Decapoda: Astacidea) through Monitoring of Invasive Species Barriers and Relocated Populations." *Journal of Crustacean Biology* 38, no. 3 (2018): 257–66.

Kawai, Tadashi. "Re-Examination of *Pacifastacus nigrescens* (Decapoda : Astacidae)." *Crustacean Research*, no. 7 (2012): 75–83.

Larson, Eric R., Magalie Castelin, Bronwyn W. Williams, Julian D. Olden, and Cathryn L. Abbott. "Phylogenetic Species Delimitation for Crayfishes of the Genus *Pacifastacus*." *PeerJ*, no. 4 (2016): 1–19.

Larson, Eric R., Cathryn L. Abott, Scott R. Gilmore, et al. "Genome Skimming Supports Two New Crayfish Species from the Genus *Pacifastacus* Bott, 1950 (Decapoda: Astacidae)." *Zootaxa* 5632, no. 3 (2025): 501–21.

Larson, Eric R., and Julian D. Olden. "The State of Crayfish in the Pacific Northwest." *Fisheries* 36, no. 2 (2011): 60–73.

Larson, Eric R., and Bronwyn W. Williams. "Historical Biogeography of *Pacifastacus* Crayfishes and Their Branchiobdellidan and Entocytherid Ectosymbionts in Western North America." In *Freshwater Crayfish*, edited by Tadashi Kawai, Zen Faulkes, and Gerhard Scholtz, 404–47. CRC, 2016.

Light, Theo, Don C. Erman, Chris Myrick, and Jay Clarke. "Decline of the Shasta Crayfish (*Pacifastacus fortis* Faxon) of Northeastern California." *Conservation Biology* 9, no. 6 (1995): 1567–77.

Pintor, Lauren M., Andrew Sih, and Marissa L. Bauer. "Differences in Aggression, Activity and Boldness between Native and Introduced Populations of an Invasive Crayfish." *Oikos* 117, no. 11 (2008): 1629–36.

US Fish and Wildlife Service. *Recovery Plan for the Shasta Crayfish (Pacifastacus fortis)*. Region 1, US Fish and Wildlife Service, 1998.


## CHAPTER 8

Foltz, David A., Nicole M. Sadecky, Greg A. Myers, et al. "*Cambarus loughmani*, a New Species of Crayfish (Decapoda: Cambaridae) Endemic to the Pre-Glacial Teays River Valley in West Virginia, USA." *Journal of Natural History* 52, no. 45–46 (2018): 2875–97.

Jezerinac, Raymond F., G. Whitney Stocker, and Donald C. Tarter. *Crayfishes (Decapoda: Cambaridae) of West Virginia*. Ohio Biological Survey, College of Biological Sciences, Ohio State University, 1995.

Loughman, Zachary J., Thomas P. Simon, and Stuart A. Welsh. "West Virginia Crayfishes (Decapoda: Cambaridae): Observations on Distribution, Natural History, and Conservation." *Northeastern Naturalist* 16, no. 2 (2009): 225–38.

Loughman, Zachary J., Stuart A. Welsh, James W. Fetzner, and Roger F. Thoma. "Conservation of Imperiled Crayfish, *Cambarus veteranus* (Decapoda: Reptantia: Cambaridae)." *Journal of Crustacean Biology* 35, no. 6 (2015): 850–60.

Loughman, Zachary J., Stuart A. Welsh, Nicole M. Sadecky, Zachary W. Dillard, and R. Katie Scott. "Environmental Covariates Associated with *Cambarus veteranus* Faxon, 1914 (Decapoda: Cambaridae), an Imperiled Appalachian Crayfish Endemic to West Virginia, USA." *Journal of Crustacean Biology* 36, no. 5 (2016): 642–48.

Loughman, Zachary J., Stuart A. Welsh, Nicole M. Sadecky, Zachary W. Dillard, and R. Katie Scott. "Evaluation of Physicochemical and Physical Habitat Associations for *Cambarus callainus* (Big Sandy Crayfish), an Imperilled Crayfish Endemic to the Central Appalachians." *Aquatic Conservation: Marine and Freshwater Ecosystems* 27, no. 4 (2017): 755–63.

Newcombe, Curtis L. "The Crayfishes of West Virginia." *Ohio Journal of Science* 29, no. 6 (1929): 267–88.

Palmer, M. A., E. S. Bernhardt, W. H. Schlesinger, et al. "Mountaintop Mining Consequences." *Science* 327, no. 5962 (2010): 148–49.

Thoma, Roger F., Zachary J. Loughman, and James W. Fetzner. "*Cambarus (Puncticambarus) callainus*, a New Species of Crayfish (Decapoda: Cambaridae) from the Big Sandy River Basin in Kentucky, Virginia, and West Virginia, USA." *Zootaxa* 3900, no. 4 (2014): 541–54.

Vopal, Christopher G., and Zachary J. Loughman. "Life History of the Big Sandy Crayfish *Cambarus callainus* Thoma, Loughman and Fetzner, 2014 (Decapoda: Astacoidea: Cambaridae), an Imperiled Crayfish in the Central Appalachian Coalfields, USA." *Journal of Crustacean Biology* 41, no. 2 (2021): 1–13.

Zipper, Carl E., and Jeff Skousen. "Coal's Legacy in Appalachia: Lands, Waters, and People." *Extractive Industries and Society* 8, no. 4 (2021): 100990.

## CHAPTER 9

Glon, Mael G., Susan B. Adams, Zachary J. Loughman, Greg A. Myers, Christopher A. Taylor, and Guenter A. Schuster. "Two New Species of Burrowing Crayfish in the Genus *Lacunicambarus* (Decapoda: Cambaridae) from Alabama and Mississippi." *Zootaxa* 4802, no. 3 (June 2020): 401–39.

Glon, Mael G., Michael B. Broe, Keith A. Crandall, et al. "Anchored Hybrid Enrichment Resolves the Phylogeny of *Lacunicambarus* Hobbs, 1969 (Decapoda: Astacidea: Cambaridae)." *Journal of Crustacean Biology* 42, no. 1 (January 2022): 1–18.

Taylor, Christopher A., Robert J. DiStefano, Eric R. Larson, and James A. Stoeckel. "Towards a Cohesive Strategy for the Conservation of the United States' Diverse and Highly Endemic Crayfish Fauna." *Hydrobiologia* 846, no. 1 (2019): 39–58.

# Index

*Page numbers in italics refer to illustrations.*

abdomen, 10, 52, 110, 191; anatomy of, *9*, *33*, *93*; as creating sound, 90–91; as food, 14; molting and, 59; reproduction and, 112–15; tail-flip escape response and, 8, *27*, *72*, *74*

Achumawi people, 162

Adams, Susie, 197

aggressive behavior, 8, 35, 70, 78, 97, 106, 140

albinism, 3

Allegheny Mountain Mudbug (*Cambarus fetzneri*), 173

all-terrain vehicles (ATVs), 180

American Lobster (*Homarus americanus*), 34

Angilletta, Michael, 69

annulus ventralis, 110, 112–13, 155

antennae, 9, 147; of *Barbicambarus*, 24; of burrowing crayfish, 31; of cave crayfish, 95, *95*; cocaine high and, 78; during communication, 100, 103, 107–8; morphology of, *94*, *101*; sensory biology and, 87, *92*, *103*

antennule, 9, 100–101, 107, *108*, 110

antidepressants, 77, 80–84

*Aphanomyces astaci*, 137

aposematism, 40

Appalachia, 4, 8, 14, 46, 148; Appalachian Mountains, 13; mining in, 168, 178–80, 182

Appalachian Brook Crayfish (*Cambarus bartonii*), 8

aquaculture, 124, 130–32, 137, 189

asexual reproduction, 138

Astacidae (family), 153, 155–56

Astacidea (infraorder), 7

Astacids. *See* Astacidae (family)

astacivores, 62

astacology, 3, 10, 34, 41, 65, 98, 130, 138, 161, 171–72, 195, 200; International Association of Astacology, 41

astaxanthin, 38

Atsugewi people, 162

Attenborough, David, 31

Australia, 9, 90, 130

*Batrochochytrium dendrobatidis*, 138

Bellman, Kirstie, 75–77

Big Creek Crayfish (*Faxonius peruncus*), 198, 199

Big Sandy Crayfish (*Cambarus callainus*), 169, 182–90, *183*, *189*

birds, 6, 38, 45, 53, 87, 90, 144, 176; as
    crayfish predators, 48, 65, 66, 67, 93;
    as crayfish transportation, 133
bladderwort, 49
blindfolded behavior, 90–92, 94, 103
Blue Crawfish (*Cambarus monongalensis*),
    29–31, 30, 36, 38–39, 43
Blue Teays Mudbug (*Cambarus loughmani*),
    173, 173
Bottlebrush Crayfish (*Barbicambarus
    cornutus*), 24, 24, 64
Bouchard, Raymond, 161
Branchiobdellidae (family), 52
Breithaupt, Thomas, 99–101
bridge, 171–72, 184–90
Brown Rat (*Rattus norvegicus*), 70
burrowing crayfish, 12, 14, 20–21, 50, 53,
    56, 80, 194, 197; behavior of, 29–41;
    classification of, 23–29; color of,
    36–44; *Distocambarus*, 28; habitat of,
    18, 19, 30; invasive species of, 121, 128,
    143; primary, 26–29, 32, 38, 44, 46, 88;
    secondary, 25, 26–27, 32; sociality of,
    31, 35, 90; tertiary, 23–27, 88; trapping
    of, 33. *See also* burrows; Hobbs
    classification
burrow plunging, 20, 22, 29, 31–33, 49,
    172, 191
burrows, 29–30, 37, 38, 40–44, 64, 74,
    80, 88, 91, 175, 185, 194, 197; aquatic,
    23–26, 93; creation of, 16; excavation
    of, 19–21, 29, 31–32, 35, 84; fossil
    burrows, 154; as habitat, 18–19, 19,
    21–22, 29, 30, 33; morphology of, 31;
    plunging of, 20, 22, 29, 31–33, 49,
    172, 191; sharing of, 34–36; as shelter,
    46–59, 84; terrestrial, 9, 12, 26–29,
    200. *See also* burrowing crayfish;
    chimneys; tunnels
burrow sharing, 34–36

Cajun, 124, 141
California, 4, 74–75, 121, 130, 146–47,
    149–53, 157, 159, 161, 163, 166
Cambaridae (family), 7, 155

Cambarids, 155
camouflage, 3, 12, 37, 40, 43
cannibalism, 36, 114–15, 160
carapace, 10, 52, 74, 80, 91, 118, 147, 156;
    color of, 120, 150, 151; fossil of, 154;
    molting and, 59; morphology of, 9,
    156
care: parental, 31, 35–36; maternal, 31,
    115
Carnegie Museum of Natural History,
    30, 173
cave crayfish, 43, 94, 95
cave habitat, 9–10, 12, 200
Center for Biological Diversity, 180–81
chimneys, 18, 19, 27, 29, 32–33, 80
Christmas Tree Crayfish (*Procambarus
    pygmaeus*), 114
Cicada Killer (*Sphecius speciosus*), 50
claws, 7, 22, 29, 34, 35, 39, 43, 49, 52,
    61–62, 64, 74, 78, 114, 116, 121, 128,
    130, 133, 135, 147, 151, 152, 174, 177,
    184; anatomy of, 8, 9, 23, 26–27,
    40–41, 105; claw loss, 152; of cave
    crayfish, 12; of crustacean, 6; mating
    and, 110, 112, 143; meral spread,
    72, 73; sensors on, 101–2; use of, in
    burrowing, 46–48; use of, in defense,
    72, 73, 90, 93; use of, in fights, 10, 97,
    103–9; when molting, 59
coalfields, 4, 169–71, 168, 174–78, 182,
    184, 190
Coalfields Crayfish (*Cambarus theepiensis*),
    176
coal fines, 169–70
coal seam, 178
color, 12, 14, 96, 129–30, 149–50, 174, 183,
    192; of burrowing crayfish, 36–44,
    200; mutations, 3; variations, 6, 14,
    42, 104, 105, 151. *See also* camouflage
color vision, 87, 90
conservation, 4, 15–16, 65, 144, 180–81,
    186, 200; of Appalachian species, 171,
    182, 184–85, 188, 190; of Shasta Cray-
    fish, 146–48, 166–67; and taxonomy,
    196–97

COVID-19, 81
Crater Lake, 117, 117, 134–35
crawfish farming, 124–28
Crawfish Frog (*Lithobates areolatus*),
    53–58, 54
Crawzilla Crawdad (*Lacunicambarus
    chimera*), 56, 56
crayfish, name origin, 5
crayfish plague, 137–38
crayfish worm, 52
crayling, 31, 64, 114, 186
crustacyanin, 38
cryptic species, 195

Dearborn, George V. N., 91
Decapoda (order), 6–7
decapods, 6–7
diet, 38, 50, 62–65, 131, 162
Digger Crayfish (*Creaserinus fodiens*),
    31–34, 32, 40, 56
dinosaurs, 16, 46
DiStefano, Bob, 65, 67
ditches, 4, 17–23, 26, 31–32, 46
Ditch Fencing Crayfish (*Faxonella
    clypeata*), 8
DNA, 194
drugs, 15, 77–78, 80–84
Dwarf Mexican Crayfish (*Cambarellus
    patzcuarensis*), 129, 129

Eastern Massasauga Rattlesnake
    (*Sistrurus catenatus*), 50
ecosystem engineers, 45, 63, 200
*écrevisse*, 5
Eger, Petra, 98–100
eggs, 55, 57, 139, 140, 143, 156, 160, 166,
    177, 186–89; of amphibian, 135–36;
    female care of, 10; laying of, 110–15;
    laying of, in crayfish burrows, 50,
    53. *See also* care; mating; ovigerous,
    female
Ellis, Maria, 146–53, 157–66
Endangered Species Act (ESA), 177–78,
    180–81, 184, 186, 190, 196, 198
Entocytheridae (family), 52

ESA. *See* Endangered Species Act (ESA)
Everglades Crayfish (*Procambarus alleni*),
    129
exoskeleton, 10, 38, 52, 58–61, 93, 97, 99,
    101, 103, 150, 170; arthropod, 6; fossil
    crayfish, 154. *See also* molting
extinction, 4, 15–16, 145, 152–53, 157, 171,
    180
eyes, 3, 37, 114, 137; anatomy of, 87, 89–90;
    blindfolding, 91; of cave crayfish, 12,
    94, 96, 200. *See also* vision

fan organs, 101
Fetzner, Jim, 17
fighting behavior, 3, 15, 35, 69–71, 74, 79,
    90, 96, 97, 104–13
fishing, 5, 14, 119, 123, 131, 132, 134, 136,
    144
fluorescein, 99
fluorescent, 99–100
fossil crayfish, 46, 153–55
Freckled Crayfish (*Cambarus maculatus*),
    41, 42
Frosted Flatwoods Salamander
    (*Ambystoma cingulatum*), 53
Fruit Fly (*Drosophila melanogaster*), 70

genetics, 6, 15, 44, 70, 139, 155, 166, 182,
    194–95
glaciers, 12
glair glands, 112–13, 188
Glon, Mael, 191–97
Glossy Crayfish Snake (*Regina rigida*), 62
golf courses, 132
Gondwanaland, 156
gonopods, 3, 1110
Graham's Crayfish Snake (*Regina
    grahamii*), 62
Great Chicago Fire of 1871, 153, 159
Great Plains Mudbug (*Lacunicambarus
    nebrascensis*), 34
Greensaddle Crayfish (*Cambarus manningi*),
    42, 43
Guyandotte River Crayfish (*Cambarus
    veteranus*), 171, 174, 174–78, 180–86, 190

Heemeyer, Jen, 57
Herberholz, Jens, 78–80
Hillbilly Hairy Crayfish (Cambarus polypilosus), 93
Hine's Emerald Dragonfly (Somatochlora hineana), 50, 51
Hobbs, Horton Holcombe, Jr., 9–12, 15, 22–29, 33, 52
Hobbs classification, 22–29
House Mouse (Mus musculus), 70
Huber, Robert, 78–79
Huxley, Thomas Henry, 14–15, 71–72, 85

Ilmawi people, 162
"in berry." See ovigerous, female
Indigenous peoples, 124, 162
indigenous species. See native species
International Association of Astacology, 41. See also astacology
invasive species, 16, 118, 147–50, 152–53, 155–57, 159–61, 176, 190, 197–201; impact of, 119–34; spread of, 123, 134–38

Jackson Prairie Crayfish (Procambarus barbiger), 103
Jacobson, Ludwig Levin, 61
Jezerinac, Raymond, 32, 172

keystone species, 45, 63, 67, 200
Kirtland's Snake (Clonophis kirtlandii), 50
K-Pg extinction event, 50
Krasne, Franklin, 75–77

Lannoo, Michael, 55–58
Laurasia, 156
Lazarchik, Andrew, 79
Least Crayfish (Cambarellus diminutus), 12, 13
leucistic, 3
Little Brown Mudbug (Lacunicambarus thomai), 21, 37, 40
lobsters, 4, 5, 6–9, 34, 38, 98, 115, 124, 128–29
Lonesome Gravedigger (Lacunicambarus mobilensis), 194

Longpincered Crayfish (Faxonius longidigitus), 104
Loughman, Zac, 171–78, 180–90

Marbled Crayfish (Procambarus virginalis), 138–40
mark-recapture, 33
Marmorkrebs, 138–40
mass extinction event, 15–16, 171
mating, 8, 10, 24, 35, 41, 43, 107, 109, 111–12, 138–39, 155
Mazama, Mount, 117
Mazama Newt (Taricha granulosa mazamae), 117, 134–35
Meadow River Mudbug (Cambarus pauleyi), 173
meral spread, 72, 73
mesocosm, 82–84
mining, 170, 178–80, 186; methods, 178; mountaintop removal, 178–80, 179
Misfortunate Crayfish (Pacifastacis malheurensis), 153
model organism, 15, 70–71, 78, 107
molting, 59–64, 114–15, 183
Moore, Paul, 103
mudbugs, 4, 21–22, 29, 34, 37, 40, 171, 194, 196–97, 200
Mud Salamander (Pseudotriton montanus), 51, 53
Murray River Crayfish (Euastacus armatus), 90–91
mussels, 178, 181

native range, 118, 133
native species, 6, 7, 12–13, 117–18, 119, 120, 121–23, 123, 133; decline of, 16, 118, 130, 132, 136–38, 144, 147. See also invasive species; nonnative species
nephropores, 98, 100–101
nervous system, 71, 74–75, 77, 79, 80
Newcombe, Curtis, 172
New River Crayfish (Cambarus chasmodactylus), 104
nonnative species, 16, 118, 123. See also invasive species; native species

Norrocky, James, 31–34
Norrocky Burrowing Crayfish Trap, 33

Okanagan Crayfish (*Pacifastacus
    okanagensis*), 153
Oregon, 117, 153–54, 159
Ortmann, Arnold, 29–30
ostracods, 11, 52
ovigerous, female, 113–14, 113, 166

*Pacifastacus chenoderma*, 154
Pacific Northwest (PNW), 146, 153
Painted Devil Crayfish (*Lacunicambarus
    ludovicianus*), 191–94, 193, 197
Pangea, 67, 155–56
parthenogenesis, 139
Pennsylvania, 2–4, 17, 19, 29, 137, 144
pet trade, 6, 10, 128–32, 138–39
pheromones, 96, 142
Piedmont Prairie Burrowing Crayfish
    (*Distocambarus crockeri*), 28
Pilose Crayfish (*Pacifastacus gambelii*), 153,
    156, 159, 161
pinch, 21, 40, 43, 49, 61, 103–8, 114, 142,
    150
Pittsburgh, 30, 36
Placid Crayfish, 152
plunging. *See* burrow plunging
PNW (Pacific Northwest), 146, 153
Prairie Crayfish (*Procambarus gracilis*), 91
primary burrowing crayfish, 26–29,
    32, 38, 44, 46, 88. *See also* secondary
    burrowing crayfish; tertiary burrowing
    crayfish
Pyblast, 144

Queen Snake (*Regina septemvittata*), 58–63

Red Claw Crayfish (*Cherax quadricarinatus*),
    130
Red Swamp Crayfish (*Procambarus clarkii*),
    7–8, 27, 70, 75, 79, 119, 121, 122,
    126–28, 133, 186
Reisinger, Alex, 82
Reisinger, Lindsey, 82

reproduction, 50, 51, 77, 111, 115; asexual,
    138; sexual, 139
reproductive cycle, 126, 186
rice, 127–28
Rock Crayfish (*Cambarus carinirostris*),
    12, 26
rostrum, 9, 149, 177, 183
Rotenone, 143
Rough-Skinned Newt (*Taricha granulosa*),
    134
Rusty Crayfish (*Faxonius rusticus*), 116, 120,
    123, 136, 136–37, 141
Rusty Gravedigger (*Lacunicambarus miltus*),
    192, 193

salamander, 2, 46, 51, 53, 96, 172, 180
Sandeman, David, 92
San Francisco, 157–59, 161
scanning electron microscope (SEM), 93
Schuster, Guenter, 41
secondary burrowing crayfish, 25,
    26–27, 32. *See also* primary burrowing
    crayfish; tertiary burrowing crayfish
seed shrimps, 11, 52, 52
selective serotonin reuptake inhibitor
    (SSRI), 83
SEM (scanning electron microscope), 93
setae, 92, 100, 102
Shasta Crayfish (*Pacifastacus fortis*), 146–53,
    151, 156–67
Shrimp Crayfish (*Faxonius lancifer*), 104
Signal Crayfish (*Pacifastacus leniusculus*),
    116–17, 121, 122, 134–35, 147–50,
    152–54, 156–57, 159–67
Skunk Cabbage (*Symplocarpus foetidus*),
    29–30
Slender Crayfish (*Faxonius compressus*),
    47, 47
Slough Crayfish (*Procambarus fallax*),
    139–40
Smithsonian National Museum of Natural
    History, 11, 153, 158
SMRT (sterile male release technique),
    143
smuggling, 130

snails, 136, 148, 152, 162, 166

Snake River Pilose Crayfish (*Pacifastacus connectens*), 153, 161, 167

snakes, 40, 49–50, 53, 57–58, 61, 61–64, 170, 173

sociality, 34–36

Sooty Crayfish (*Pacifastacus nigrescens*), 153, 157–61

Southern Cave Crayfish (*Orconectes australis*), 43

species complex, 195

species status assessment (SSA), 181–82

Spider Cave Crayfish (*Troglocambarus maclanei*), 95, 95

Spinycheek Crayfish (*Faxonius limosus*), 82, 82

Spiny Stream Crayfish (*Faxonius cristavarius*), 175

Spring Rivers, 147–48, 157

springs, 146–50, 157, 161–65

SSA (species status assessment), 181–82

SSRI (selective serotonin reuptake inhibitor), 83

sterile male release technique (SMRT), 143

St. Francis River Crayfish (*Faxonius quadruncus*), 198, 199

Stocker, Whitney, 172

stream banks, 2, 25, 26, 46, 169, 176, 185, 187–88

Striped Crayfish Snake (*Liodytes alleni*), 62

sundews, 49

Supernova Crayfish (*Cherax boesemani*), 129

Swierzbinski, Matthew, 79

tail fan, 9, 74, 114

tail-flip escape response, 27, 73, 74–76, 79–80, 109, 163

Tasmanian Giant Freshwater Crayfish (*Astacopsis gouldi*), 65

taxonomy, 7, 11, 173, 194, 197–98

territoriality, 34, 92, 106–7, 147, 157, 160

tertiary burrowing crayfish, 23–27, 88. *See also* primary burrowing crayfish; secondary burrowing crayfish

Texas Prairie Crayfish (*Fallicambarus devastator*), 41

Thoma, Roger, 172, 181

Thunderbolt Crayfish (*Cherax pulcher*), 129

translocation, 186–90

Tug Valley Crayfish (*Cambarus hatfieldi*), 173

tunnels, 19–21, 26, 31, 46–47, 49, 56, 58

umwelt, 85–88, 90–91, 95–96, 103, 107

Upland Burrowing Crayfish (*Cambarus dubius*), 14

urine, 3, 81, 86, 87, 95–103, 106, 108–9

US Fish and Wildlife Service (USFWS), 163, 177, 180, 181–82, 184, 186–87

USFWS. *See* US Fish and Wildlife Service (USFWS)

Valley Flame Crayfish (*Cambarus deweesae*), 38–39

Virile Crayfish (*Faxonius virilis*), 115, 120, 121, 123, 132, 160

vision, 37, 75, 87–91, 94–96, 103; color vision, 87, 90. *See also* eyes

von Uexküll, Jakob Johann, 87

water mold, 137

water quality, 14, 184

West Liberty University, 172, 184–90

West Virginia, 29, 169- 186

White Sulphur Springs, 186, 188, 189–90

Wiersma, Cornelis A. G., 74

Wizard Island, 116, 117

Woodland Crayfish (*Faxonius hylas*), 198

Yabby (*Cherax destructor*), 128

Zebra Fish (*Danio rerio*), 70, 72